T0283657

ELECTRONIC THEFT

Unlawful Acquisition in Cyberspace

The convergence of communications and computing has begun to transform Western industrial societies. Increasing connectivity is accompanied by unprecedented opportunities for crimes of acquisition. The fundamental principle of criminology is that crime follows opportunity, and opportunities for theft abound in the digital age. *Electronic Theft* names, describes and analyses the range of electronic and digital theft, and constitutes the first major survey of the field. The authors cover a broad list of electronic misdemeanours, including extortion, defrauding governments, telephone fraud, securities fraud, deceptive advertising and other business practices, industrial espionage, intellectual property crimes, and the misappropriation and unauthorized use of personal information. They have been able to capture impressively large amounts of data internationally from both scholarly and professional sources. The book poses and attempts to answer some pressing questions to do with national sovereignty and enforceability of laws.

Peter Grabosky is Director of Research at the Australian Institute of Criminology and President of the Australian and New Zealand Society of Criminology. He is the author of numerous books, including *Of Matters Gentle: Enforcement Strategies of Australian Business Regulatory Agencies* (with John Braithwaite, 1987).

Russell G. Smith is a senior research analyst with the Australian Institute of Criminology and the author of several books on crime and the health care professions. He is co-author (with Peter Grabosky) of *Crime in the Digital Age: Controlling Telecommunications and Cyberspace Illegalities* (1998).

Gillian Dempsey is a barrister and a senior lecturer in law at the University of Queensland, specialising in information technology and intellectual property law.

ELECTRONIC THEFT

Unlawful Acquisition in Cyberspace

PETER GRABOSKY
Australian Institute of Criminology

RUSSELL G. SMITH
Australian Institute of Criminology

GILLIAN DEMPSEY
University of Queensland

CAMBRIDGE
UNIVERSITY PRESS

477 Williamstown Road, Port Melbourne, VIC 3207, Australia

Cambridge University Press is part of the University of Cambridge.

It furthers the University's mission by disseminating knowledge in the pursuit of education, learning and research at the highest international levels of excellence.

www.cambridge.org
Information on this title: www.cambridge.org/9780521805971

First published 2001
Reprinted 2002
First paperback printing 2010

A catalogue record for this publication is available from the British Library

National Library of Australia Cataloguing in Publication data
Grabosky, Peter N. (Peter Nils), 1945–
Electronic theft: unlawful acquisition in cyberspace.
Bibliography.
Includes index.
ISBN 0 521 80597 X.
1. Computer crimes. I. Smith, Russell G.
II. Dempsey, Gillian. III. Title.
364.168

ISBN 978-0-521-80597-1 Hardback
ISBN 978-0-521-15286-0 Paperback

Contents

Preface

As with most books, this could not have been written without the assistance of many hands.

We are grateful to the Australian Institute of Criminology for institutional support during the writing of this book, and to the Institute's J. V. Barry Library for its outstanding assistance throughout the life of the project. Various Australian police services provided helpful insights into the difficulties of investigating electronic fraud.

We wish to thank Ajoy Ghosh for reviewing a draft of the entire work, and Andrew Greinke for his reading of a number of chapters. Dr Robert O'Connor, Katy Roberts, Dr Gregor Urbas, and Bernard Wall provided excellent (and timely) research assistance.

Some of our ideas relating to legal pluralism have been developed from our earlier work: P. N. Grabosky and Russell G. Smith, 'Telecommunications and Crime: Regulatory Dilemmas', *Law and Policy* 19(3) 1997: 317–41; and Peter Grabosky, 'Using Non-Governmental Resources to Foster Regulatory Compliance', *Governance: an International Journal of Policy and Administration*, 8(4) 1995: 527–50.

We are grateful to our publishers, Cambridge University Press, for their guidance; to three anonymous reviewers for their generous and helpful advice; and to the long-suffering participants at the many conferences we addressed who offered helpful comments and observations on earlier drafts of our chapters.

A special note of thanks goes to Venetia Somerset for her superb editorial skills.

Any errors or shortcomings remain the authors' joint responsibility.

Abbreviations

ABA	Australian Broadcasting Authority
ACCC	Australian Competition and Consumer Commission
ACT	Australian Capital Territory
ADMA	Australian Direct Marketing Association
AICPA	American Institute of Certified Public Accountants
AOL	America OnLine
ASIC	Australian Securities and Investment Commission
ASX	Australian Stock Exchange
ATM	automated teller machine
CAUDIT	Committee of Australian University Directors of Information Technology
CD	compact disc
CERTS	computer emergency response teams
CHAPS	Clearing House Automated Payment System
CIA	Central Intelligence Agency
CNN	Cable News Network
DCTIA	Department of Communications, Information Technology and the Arts
DPRL	Digital Property Rights Language
DVD	digital video disc
EBT	electronic benefits transfer
EFT	electronic funds transfer
EFTPOS	electronic funds transfer at point of sale
EMR	electromagnetic radiation
FAQ	frequently asked question
FBI	Federal Bureau of Investigation
FSA	Financial Services Authority (UK)
FTC	Federal Trade Commission (USA)
GST	goods and services tax
HIC	Health Insurance Commission

ICAC	Independent Commission Against Corruption
ICC	International Chamber of Commerce
ICQ	'I Seek You'
IMSN	International Marketing Supervision Network
IOSCO	International Organization of Securities Commissions
ISP	Internet service provider
IT	information technology
ITAA	Information Technology Association of America
JPEG	Joint Photographic Experts Group
KPMG	Klynveld Peat Marwick Goerdeler
MPEG	Moving Pictures Expert Group
NASAA	North American Securities Administrators Association
NASD	National Association of Securities Dealers (USA)
NASDAQ	National Association of Securities Dealers Automated Quotation (system)
NFIC	National Fraud Information Centre
NSW	New South Wales
NYSE	New York Stock Exchange
OECD	Organization for Economic Cooperation and Development
P3P	Platform for Privacy Preferences
PIN	personal identification number
RIAA	Recording Industry Association of America
SDMI	Secure Digital Music Initiative
SEC	Securities and Exchange Commission (USA)
SEXI	Systems of Excellence Inc.
SOMA	System for Surveillance of Market Activity
UNCITRAL	United Nations Commission on International Trade Law
VAT	value-added tax
WAP	Wireless Application Protocol
WIPO	World Intellectual Property Organization

CHAPTER 1

Theft and Cyberspace

The convergence of communications and computing has begun to transform the world. Although this transformation has mainly been located in Western industrial societies, its effects are beginning to be felt globally. As recently as 1990, few people could even envisage what has since come to be known as the World Wide Web, and the possibility that an ordinary individual could communicate with millions of others seemed remote. More traditional telecommunications offered little assistance – without 'the phone book', obtaining a person's telephone number necessitated a laborious, often time-consuming, and increasingly costly call to Directory Assistance. Who would have imagined a desktop link to an electronic white pages, searchable by typing in the name or address of the subject? The idea of making near instantaneous copies of vast amounts of text, much less video, sound and multimedia, was the stuff of science fiction. Our ability to gather and to disseminate information now vastly exceeds our capacity to absorb and to analyse it.

The enormous increase in human potential brought about by the democratization of digital technology, however, has its downside. Information is now being disseminated in such volume that understanding and control by any individual is implausible. Today cyberspace is perceived to be teeming with advertising, vice and (for the time being at least) American content, and critics all over the terrestrial world lament the erosion of traditional values they consider worth keeping.[1]

Beyond its contribution to culture wars, increasing connectivity has been, and will continue to be, accompanied by unprecedented opportunities for crimes of acquisition. The extent to which business transactions are being conducted electronically is increasing enormously. Forrester Research estimated that global business-to-business online commerce could amount to US$2.7 trillion by the year 2004, while the Gartner Group puts the figure closer to US$7 trillion in 2004 (*San Jose Mercury News,* cited in O'Brien 2000). One may expect the growth of electronic commerce to be reflected in the growth of electronic misappropriation. The fundamental principle of criminology is that crime follows opportunity, and opportunities for theft abound in the Digital Age.

This book will review some of the major crimes of acquisition involving digital technology as the instrument of theft, or crimes involving information as the object of theft. Many of these crimes are not 'new' – rather it is the medium of theft, the transjurisdictional reach of the thief, and the speed with which a transaction may be executed, that are without precedent. Chapters 8 to 10 also discuss some forms of misappropriation whose legal status is contested or evolving, such as the commercial acquisition of personal information, industrial espionage, and various activities relating to digital piracy.

The Criminology of Electronic Theft

Cohen and Felson (1979) observed that all crime can be explained by the intersection of three variables: a supply of motivated offenders, the availability of suitable targets, and the absence of capable guardians – someone to mind the store, so to speak. Crime reduction strategies may be directed at each of these factors: reducing an offender's motivation through moral suasion, or failing that, the deterrent effects of prosecution and punishment; making it more difficult to offend through target hardening; and increasing surveillance in order to enhance the possibility that illegal activities will be detected.

These basic principles of criminology apply to computer-related crime no less than to motor vehicle theft or to shoplifting, and they will be evident throughout this book. As we will note, not all of these factors are amenable to control by governments alone. It follows, therefore, that a variety of institutions will be required to control computer-related crime.

The motives of those who would commit computer-related crime are diverse, but hardly new. Computer criminals are driven by time-honoured motives, the most obvious of which are greed, lust, power, revenge, adventure, and the desire to taste 'forbidden fruit'. The ability to make an impact on large systems may, as an act of power, be gratifying in and of itself. The desire to inflict loss or damage on another may also spring from revenge, as when a disgruntled employee shuts down an employer's computer system, or from ideology, as when one defaces the web page of an institution that one regards as abhorrent. Much activity on the electronic frontier entails an element of adventure – the exploration of the unknown. The very fact that some activities in cyberspace are likely to elicit official condemnation is enough to attract the defiant, the rebellious, or the irresistibly curious. Given the degree of technical competence required to commit many computer-related crimes, there is one other motivational dimension worth noting here. This, of course, is the intellectual challenge of mastering complex systems. None of the above motives is new. An element of novelty, however, resides in the unprecedented capacity of technology to facilitate the ability of individuals to act on these motives. Our focus in this book is on theft (in the general sense of unlawful acquisition), so the first of these will figure prominently in the chapters that follow. But there is often more to theft than mere greed, and the reader will notice that many of the illegalities discussed below flow from a variety of motives.

While motives tend not to change, the variety and number of opportunities for cybercrime are proliferating. The exponential growth in connectivity of computing and communications, and their applications to electronic commerce, increase both the number of prospective victims of computer-related crime and the number of prospective offenders. As the Internet becomes increasingly a medium of commerce, it will also become increasingly a medium of fraud.

The third basic factor that explains computer-related crime is the absence of a capable guardian. Capable guardianship has evolved over human history, from feudalism, to the rise of the state and the proliferation of public institutions of social control, to the postmodern era in which employees of private security services vastly outnumber sworn police officers in many industrial democracies. Here again, it may be instructive to compare computer-related crime with more conventional types of crime.

Guardianship against conventional crime involves preventive efforts on the part of prospective victims, contributions by members of the general public or commercial third parties (such as insurance companies and private security services), and the activities of law enforcement agencies. Indeed, it is often only when private efforts at crime prevention fail that the criminal process is mobilized. So it is that owners of motor vehicles are encouraged to lock their vehicles at all times, that insurance contracts may offer premium discounts for crime prevention measures such as theft alarms, and that some car parks have video surveillance or private security guards in attendance. Often it is only when these systems fail that the assistance of law enforcement is sought. So it is in cyberspace as well.

Law and Technology

It has become trite to suggest that law and policy often fail to keep pace with technological change. The most robust legal systems are those that can adapt to changing technological circumstances without continuing repeal and reenactment of legislation – so-called technology-neutral laws. By contrast, in some jurisdictions around the world, the existing law of theft has only limited application to intangible property. For instance, at common law in both England and Australia, larceny is the unlawful taking or conversion of anything capable of being stolen (Fisse 1990: 197). Because larceny could only be committed in respect of something capable of being asported (physically removed), the intangible nature of items such as telephone services rendered successful prosecution as larceny impossible. For instance, in *Low v. Blease*

([1975], 119 Sol J 695 Crim LR 513), the English Queen's Bench Divisional Court held that electricity does not constitute property within the meaning of s. 4 Theft Act 1968 (Eng). Accordingly, electricity could not be stolen by switching on the current.

This definitional problem has largely been resolved in Australia by the introduction of various statutory definitions of theft which include intangible items of property such as electricity, thus bringing the unauthorized use of a telephone service within the definition of theft – at least of the electricity consumed (in Australia, see *Crimes Act* 1900 [NSW] s. 154C; *Criminal Law Consolidation Act* 1935 [SA] s. 154; Queensland Criminal Code s. 408; Western Australia Criminal Code s. 390; Tasmanian Criminal Code s. 233). But the value of the electricity used in a telephone call is not necessarily representative of the value of the service being provided, so more specific criminal regulation is required.

Protection for intangible property, however, has typically centred on laws proscribing the obtaining of various forms of advantage by deception. For example, in Australia, section 29A of the *Crimes Act* 1914 (Cwlth) proscribes (with a maximum sentence of ten years) obtaining from the government with intent to defraud or by false pretence 'any chattel, money, valuable security or benefit'. The use of such laws to prosecute digital theft is by no means simple, however. Prosecuting larceny in the area of computer crime is complicated by the necessity of proving that the property was taken without the victim's consent. Where a computer alone, rather than a human actor, has been deceived, this lack of consent can be difficult to establish (see Grabosky and Smith 1998). In the Australian case of *Kennison* v *Daire* ((1986) 160 CLR 129, 20 February 1986), the High Court of Australia upheld the conviction of a man who had exploited a loophole in the Savings Bank of South Australia's automated teller machine (ATM) system. Kennison had closed his account with the bank but retained his card, one of the conditions of use of which was that his account was in sufficient credit to accommodate a withdrawal. The ATM had been programmed to permit up to $200 to be withdrawn by any person using a card and its correct personal identification number (PIN). Knowing his account to be empty, he used the card while the ATM was not connected to the bank's central computer. The conviction was upheld because Kennison had acted fraudulently with intent permanently to deprive the bank of $200. Kennison's argument, that the bank had consented to the withdrawal of the money by providing a machine that permitted the transaction to take place, was rejected.

To avoid any potential for misunderstanding, some risk-averse jurisdictions now include deception relating to a computer system expressly within the terms of financial crimes. For example, section 17.1 of the Australian Model Criminal Code defines deception as 'conduct by a person that causes a computer system or any machine to make a response that the person is not authorised to cause it to do'.

The chapters that follow examine a wide variety of circumstances in which benefits are able to be obtained illegally and improperly by individuals making use of digital technologies. Some clearly fall within traditional conceptions of theft; others may give rise to civil or administrative consequences, while others again may have primary relevance to personnel management in the workplace. Using a government computer for personal use during working hours might, for example, technically amount to theft, but it is unlikely that a criminal prosecution would be undertaken; rather, the matter would probably be dealt with internally. Transferring $1 million from a business bank account to a private individual's personal account, however, would usually (though not invariably) result in the matter being reported to the police and a prosecution ensuing. Economic crime in cyberspace, therefore, comprises a diverse range of behaviour with a variety of legal and social consequences.

Theoretical Context

It has long been recognized that the criminal justice system is a very imperfect means of social control, and that effective crime prevention requires the contribution of families, schools, and many other institutions of civil society. Moreover, as environmental criminologists and exponents of situational crime prevention will attest, the design of public space and the 'engineering out' of criminal opportunities can make a significant contribution to crime control.

The same principles apply to white-collar and sophisticated crime. It is now also increasingly acknowledged that effective business regulation is beyond the capacity of governments alone and requires the involvement of many other institutions. Two decades ago, Bardach and Kagan (1982: 33) maintained that most regulation was already 'in the hands not of government officials but of the myriad individuals employed in the private sector'. Teubner (1983) has spoken of how, instead of imposing direct substantive controls on behaviour, states might structure mechanisms for self-regulation. He refers to the fostering of a 'regulated autonomy' based largely on private orderings. The state facilitates the development of these self-regulating systems rather than engaging in direct intervention. Indeed, in some areas the state plays an even more passive role, deferring to private institutions. A decade later, Ayres and Braithwaite (1992) envisaged a tripartite regulatory system, embracing monitoring by government agencies, self-regulation by companies and industry associations, and surveillance and lobbying by public interest groups. These three types of institution would play an active regulatory role, strengthening the self-regulatory initiatives of regulated industries and complementing the activity of government regulatory agencies. Shearing (1993) views regulatory systems in an even wider perspective, speaking of 'regulatory space' as comprising a variety of institutional orderings and regulatory mechanisms. Grabosky (1994, 1995a) has discussed the quasi-regulatory activities of commercial institutions, observing that in certain settings the influence they have wielded was far in excess of what government agencies were able or willing to mobilize. In some instances, resources outside the public sector may be consciously harnessed by government in furtherance of regulatory objectives. In other cases, institutional orderings may arise more or less spontaneously, but with significant regulatory effects. The prevention of crimes of acquisition in cyberspace provides one of the best illustrations of how this diversity of regulatory activity may function.

This notion of legal pluralism provides the theoretical basis for this book. It is a perspective that sees law as having its analogues in various other social institutions, or where official and unofficial forms of ordering coexist in an interactive relationship (Merry 1988). Observers as long ago as Ehrlich (1912) began to see law as but the top stratum from which a web of quasi-legal rules and controls ordered everyday life. More recent scholars began to focus on the relationship between state law and private forms of social control and conflict resolution (Fitzpatrick 1984). Today there is growing realization that the capacity of the state to control both individual and corporate behaviour has limits. The nature of digital technology, combined with the ambivalence of many governments about the extent of their role in the economy and society of the twenty-first century, makes legal pluralism a useful lens for viewing the ordering of cyberspace.

One of the seminal thinkers of the late twentieth century, Michel Foucault, made a number of perceptive observations about government and society, two of which are especially germane to this book. Although he made them before the world entered the digital age, they apply no less to matters of social control in cyberspace than they do to terrestrial affairs. The first of these notes the trend towards a less central role for the state and for law in the ordering of relationships. Foucault sees overt enforcement of law as but one element in what might be called a web of constraint, some strands of which are barely discernible, and many of which are non-governmental. He refers to law as 'partial' (1980: 141) and 'not what is important' (1979: 13), observing that the real practice of government was not the imposition of law but rather working with and through the constellation of interests, institutions and interpersonal relations that are part of civil society. Burchell (1991: 127) adds further texture to Foucault's outline when he refers to 'governing in accordance with the grain of things'. Foucault saw power in modern society no longer centralized in the Sovereign, but rather dispersed. He uses the terms 'capillary power' and 'micro-physics of power' (Foucault 1977: 139) as metaphors for the partial displacement of law by other orderings. At the same time, Foucault acknowledges that law is not isolated from these microstructures of power but rather interacts with them. In this way he echoes Donald Black's (1984) theory that law varies inversely with other forms of social control.

The second of Foucault's observations which is of concern to us here is the relationship between knowledge and power. His celebrated work *Discipline and Punish* (an unfortunate translation of the original *Surveiller et Punir*) (1977) describes the transition from physical punishment as an expression of political power to surveillance and control through observation, record-keeping, monitoring and inspection. These two basic ideas converge. Not only are the surveillance capabilities of governments in Western industrial societies without precedent, but non-state actors, from large multinationals to insurance companies to parents, are all able to command formidable knowledge, and thereby control, over individuals. Chapter 10 of this book will cite some examples of the knowledge and power held by commercial interests.

So it is that states often seek to co-opt private institutions in furtherance of law enforcement and regulatory objectives. The assistance of Internet service providers (ISPs) can be helpful, or even essential to the investigation of computer-related crime. At the same time, we see private institutions seeking a market niche. A vast information security industry has grown up to complement, if not largely eclipse, whatever role government law enforcement and regulatory agents may wish to play. Rose and Miller (1992) refer to this as 'governing at a distance', using 'new technologies of government' that harness energies residing outside the public sector to advance public policy.

This line of argument has been popularized in the very influential North American book *Reinventing Government* (Osborne and Gaebler 1992). Recognizing that government as traditionally configured has its constraints and limitations, the authors advocate that governments adopt the role of facilitator and broker rather than that of commander. They suggest that governments 'steer' rather than 'row', and that they structure the marketplace so that naturally occurring private activity may assist in furthering public policy objectives. Osborne and Gaebler (1992: 280) use the term 'leverage' to refer to this approach.

The notion of legal pluralism is very salient in the digital age, where the limits of the state have become increasingly apparent. Throughout this book, we will see examples of non-state institutions and market forces exercising power no less coercive than that available to governments.

Lessig argues that behaviour in cyberspace is already controlled primarily not through legislation but through other codes: the programming and architecture of information systems. To Lessig, technology, not law, is the predominant regulatory institution of cyberspace. 'Code can, and increasingly will, displace law as the primary defense of intellectual property in cyberspace.' His emphasis is on 'private fences, not public law' (1999: 126). He observes a shift in effective regulatory power 'from law to code, from sovereigns to software' (1999: 207).

Just how 'code functions as law' is illustrated by the unique identifier that is created with every MS Word document, and by digital content (text, sound, visual, or multimedia), which can be programmed to degrade if and when it is copied. Lessig (1999: 160) contends that protection of privacy will be achieved through systems that will allow users to 'articulate their preferences and negotiate the use of data about them'. Similarly, Rochlin (1997: 13) refers to the 'hegemony of design' as constituted by 'new modes of organizational control, indirect and diffused', disguising 'the imperatives of compliance as no more than a set of operating rules'.

Much as Molière's Bourgeois Gentilhomme, who remarked that he had been speaking prose all of his life without realizing it, Lessig and Rochlin speak the language of legal pluralism. Code, market forces, and to a lesser extent social norms, have eclipsed law as the major institutions of social control in cyberspace. As Foucault did, they regard the relationship between these various elements as interactive.

Transformation of Social Relations

It was not that long ago, even in the industrialized world, that commerce was largely based on personal relationships, deals were sealed with a handshake, and 'a man's word was his bond'. The Internet has indeed brought about significant changes in human interaction. Instead of a face-to-face relationship with our local bookseller, we are just as likely to deal with a digital bookseller such as Amazon.com. But trust is the foundation of commerce in cyberspace, just as it is on the ground. The difference is that whereas in real life trust is based on personal relationships, online trust is based on confidence in processes. The establishment of trusted processes in cyberspace is the key to commercial success, and so it is that online merchants seek to create an environment in which a prospective customer can be relaxed and confident about any prospective transactions.

Although commercial transactions in cyberspace are completely depersonalized in the sense that one does not deal directly with a sentient being (other than in some real-time interactive environments), technology now enables an online bookseller to 'remember' precisely what books you have been buying, at what price, and when. This may be advantageous, as when you are automatically advised of new publications on a topic that appears to have interested you in the past, but it may be less so if information is disseminated to other merchants, to government agents, or to anyone else with an interest in your private life.

As we will see in Chapter 6, ordinary investors are now able to buy and sell shares online without dealing through intermediaries such as underwriters, brokers and investment advisers. While this may enhance the efficiency of securities markets, it also provides opportunities for criminal exploitation. But the fundamental criminality is still reducible to the basics: misrepresenting the underlying value of a security at the time of the initial public offering; or market manipulation during secondary trading of a security, through the dissemination of false information, or by engineering a deceptive pattern of transactions to attract the attention of the unwitting investor.

Computing and communications technologies make such activities both easy to carry out and difficult to detect.

The issue of trust surfaces again in the context of electronic cash, and the possibility of 'private money'. Before the nationalization of currencies a century ago, many individual banks issued their own notes. Banks differed in terms of their viability. Some issued currency that turned out to be worthless, while others' paper was 'as good as gold'. To the extent that the nature of money changes in the digital age, trust in private institutions will become even more salient.

The Limits of Government

Even where the law, through interpretation or amendment, comes close to keeping abreast of both technology and the 'disintermediation' that characterizes new commercial relationships, its use may be limited. Despite the fact that new technologies allow us to do more things more quickly than ever before, they pose challenges to governments which are not unlike those that plague many other areas of social and political life.

Perhaps the most prominent of these entails the limits of the law or indeed, government efforts generally, as instruments of public policy. To be sure, digital technologies enable a degree of surveillance that would do George Orwell proud. But not all states today are preoccupied with social control, either in cyberspace or on the ground. Those that are often realize that control can be gained or kept at a prohibitively high cost in terms of the contentment and creativity of the public and the health and vibrancy of the economy.

As we will see throughout the chapters that follow, many governments lack the capacity to control behaviour in cyberspace, on the part of their own citizens or those of other countries. This is also reflected in the limited ability of governments alone to protect personal property in the digital environment. Law enforcement resources are constrained, and there are relatively few skilled computer crime investigators and forensic accountants. The rarity of suitably trained investigators generates competition for scarce investigative resources. This effectively forces a choice between concentrating efforts on protecting property interests or focusing on areas that have attracted greater concern from the general public such as Internet child pornography and the information to facilitate the illicit production of drugs or explosives.

As far as traditional property crime is concerned, there are, in terrestrial space as well as cyberspace, very real limits to what the state can do to protect a citizen's assets. In English-speaking societies at the very least, the capacity of public police is now acknowledged to be limited. Most victims of residential burglary are aware that they stand little chance of recovering their lost possessions; they harbour few illusions that 'their' offender will eventually be brought to justice. The role of the police is often limited to that of legitimizing insurance claims and providing a few kind words (and perhaps some crime prevention advice) to the victim. Individuals are, to an extent that few wish to acknowledge openly, their own best chance of crime prevention. The result is that those who can afford the expenditure tend to acquire sophisticated alarm systems, live in 'gated' communities, and fit their motor vehicles with sophisticated locking and disabling devices.

Compounding this is the global nature of cyberspace, where crimes may be committed over vast distances and across national frontiers. Although it may be legally possible to prosecute offenders who are alleged to have committed electronic theft in various jurisdictions, even internationally, significant problems may arise in detecting such illegality and in proving allegations successfully. As long ago as 1986, an OECD paper discussing the problems of prosecuting international computer crime observed:

> For international cooperation to be effective there must be agreement as to what is criminal at the national level and what sanctions should attach to a given offence. Extradition treaties need to be adequate to deal with offences committed in various jurisdictions. There are also problems of differing laws as to search and seizure, service of documents and the taking of testimony or statements of persons. (OECD 1986: 68)

These issues are no less problematic today than they were at the dawn of the digital age. The legal challenges lie not so much in attempting to create uniform legislation internationally, which is unlikely ever to occur, but rather in ensuring that individual countries are able to prosecute those offenders and offences that occur within their own geographical boundaries.

The necessity of self-reliance in crime control is no less in cyberspace than in one's terrestrial neighbourhood. This raises the fundamental issue of what forms of cyber-theft are sufficiently threatening to warrant the full application of the criminal law, and which might be more effectively and efficiently controlled by prevention or by conferring enforceable civil property rights on the 'owner'.

New Instruments of Social Control in the Digital Age

Just as technology has brought about unprecedented opportunities for crime, so it has wrought new means of preventing and controlling crime. Security technology, including mechanisms of access control and authentication, are the equivalent of deadlocks and back-to-base alarm systems for safeguarding terrestrial property. It is these technologies of self-help, rather than any state presence, that will provide the basis for secure electronic commerce.

As an alternative to conventional law enforcement, governments may seek to confer entitlements and allow citizens to enforce their own rights. Private redress is of course not a perfect solution. Within a developed society, access to justice is unevenly distributed. Microsoft can protect its interests much more effectively than most private individuals can protect their interests. The challenge is even greater in the global context, where 'electronic pirates' can operate from jurisdictions whose authorities may be unconcerned with rights to the intellectual property of copyright-holders on the other side of the world, or may lack the capacity to enforce them.

The digital age has dramatically energized the debate on the proper balance between the public interest and private interests. While issues of content regulation and telecommunications interception are perhaps most vivid, there is also the question of when the wider public interest should require state intervention into what would otherwise be a private matter. The degree to which the state should intervene to protect digital private property in cyberspace seems likely to remain a contested issue for some time to come.

Structure of the Book

This introductory chapter explores some of the main themes of the book. The following nine chapters explore specific forms of misappropriation using digital technology. Chapter 2 provides an overview of electronic payments systems and their vulnerability to criminal exploitation. It reviews the major media of electronic payments, including direct debit, EFT, home banking systems, card-based systems, and electronic cash. Given the enormous profits that are certain to flow from secure electronic commerce, it is not surprising that the identification of security flaws in electronic commerce systems is a major goal of both honest and dishonest players, albeit for very different reasons. In this, as in most other areas of electronic theft, preventive strategies will remain the first line of defence.

Chapter 3 looks at extortion, hardly a new crime but one in which some elements have changed dramatically with the advent of digital technology. An extortion threat to damage information systems, or to disseminate embarrassing information, can now be carried out from the other side of the globe. One of the most difficult challenges faced by traditional extortionists is that of obtaining and securing an extortion payment without divulging one's identity or placing oneself at risk of arrest. Technologies of funds transfer now make this much easier, at least where the extortionist is physically situated in a jurisdiction with financial secrecy laws, and/or a prevailing official disinclination to cooperate in the investigation of transnational crime.

Chapter 4 provides an overview of the means by which governments can be digitally defrauded. The increasing presence of government activity online is matched by opportunities to exploit government's systems for illicit financial gain. Government benefit systems such as social security and health insurance are increasingly the subject of manipulation by individuals seeking to obtain more than their legal entitlement. Salary systems may be similarly attractive as targets, and, in the case of taxation systems, including customs and excise fraud, some people will seek to evade payment either in part or altogether. These various threats to the financial integrity of governments, and additional abuses such as the personal use of government technology, require the introduction of various new control systems, most of which are themselves digitally based. The chapter concludes with an overview of prevention strategies, many of which are applicable to information systems generally, whether in the public or private sectors.

Chapter 5 looks at telecommunications fraud, including misappropriation of telephone and Internet services. In its most common form, this tends to be 'low-tech' in nature – the fraudster simply fails to pay the bill and disappears. More complicated misappropriations involving sophisticated forms of fraud are exemplified in the activities of 'phreakers', the term commonly used to refer to those who seek to obtain telephone services without incurring a fee, by manipulating systems electronically. In response to these risks, many telecommunications carriers and service providers have developed sophisticated security measures, such as software and skilled investigative capacity. The proliferation of carriers and service providers in the deregulated telecommunications environments of most Western industrial societies has been accompanied by increasing cooperation within the industry on matters related to security and investigation. Indeed, because of the widely acknowledged difficulty in prosecuting telecommunications fraudsters, many matters never reach police attention and are handled through the adoption of other measures. As will be seen in other chapters, this problem is by no means unique to theft of services.

Chapter 6 describes how securities markets have changed in the digital age. Here the most dramatic developments are those that have seen the decline of intermediaries – brokers, underwriters, researchers and analysts, who traditionally (albeit sometimes imperfectly) provided some safeguard against misrepresentation when shares are initially offered to the public, or subsequently traded on stock exchanges. Today, investors have instant access to an unprecedented volume of information, and can execute their own trades directly, even from a mobile communications device. Not all of this information is accurate or truthful, however, and some may be fraudulent. The challenge to investors is that of quality control, distinguishing useful information from hype or from blatant falsehood. The challenge to regulatory authorities, whose goal is to preserve the integrity of capital markets and to maintain public confidence in them, is to identify misconduct quickly and neutralize it effectively. This will entail new methods of surveillance, including a great deal of websurfing by regulators and the application of digital technology to investor education.

Chapter 7 discusses deceptive and misleading online advertising and business practices. Many traditional frauds, such as pyramid schemes and endless variations on advance-fee scams, lend themselves easily to the digital environment. So too does the non-delivery of products, or the delivery of defective merchandise. As is the case with other forms of electronic theft, these can be committed from the other side of the globe as easily as from next door, making investigation and prosecution that much more difficult.

The globalization of deceptive advertising has begun to inspire international cooperation in enforcement and consumer education, a promising complement to traditional enforcement solutions. In addition, the online services industry is well placed to help foster electronic fair trading.

Chapter 8 explores the theft of intellectual property. Digital technology permits effectively perfect reproduction and rapid dissemination of print, graphics, sound, and multimedia combinations. The temptation to reproduce copyrighted material for personal use, for sale at a lower price or even for free distribution, has proved irresistible to many. Whether this denies artists their due and chills creativity is an important question. There are those who maintain that intellectual property should be accorded the same legal protection as physical property, while at the other extreme there are those who argue that all information should be free. Meanwhile, technologies such as digital watermarks have been developed which protect against unauthorized copying. Technologies based on unique identifiers that leave a trail of evidence if unauthorized copying occurs may reassure the original owner, but the idea that the consumer's reading, viewing or listening patterns may be recorded, and ultimately retrievable, raises problems that are addressed more fully in the subsequent chapter on misappropriation of personal information.

Chapter 9 deals with industrial espionage. Increasing connectivity places companies at greater risk of losing proprietary information such as trade secrets, as well as strategic intelligence. This risk is compounded by the widespread availability of interception technologies, including hacking tools and spying instruments such as directional laser listening devices. Moreover, changes in organizational life have seen many companies now experiencing significant personnel turnover. 'Insiders' have traditionally constituted a greater threat to an organization's assets than have 'outsiders', and the risk of betrayal, whether deliberate or accidental, is greater in a fluid labour market. Responses to industrial espionage range from harsh criminal sanctions with the assertion of extraterritorial jurisdiction, such as that which characterizes the United States, to reliance on private civil remedies and security measures in those jurisdictions where the prevailing philosophy and the capacity of the state may not permit such aggressive action by government.

Chapter 10 looks at the electronic exploitation of personal information. While governments have traditionally been regarded as the greatest threat to individual privacy, new technologies have now provided private interests with the capacity to acquire and collate vast amounts of information on one's personal life. While this can facilitate customer service in the new world of e-commerce, misuse of this information for purposes of stalking, harassment, intimidation or manipulation is not uncommon. As is the case with many computer-related mishaps, a great deal of private information can also be disclosed by accident. This raises the question of whether the protection of personal information would best be achieved by conferring property rights and allowing market forces to prevail. In the short term, many consumers will be attracted to those areas of cyberspace that promise the protection of privacy.

The concluding chapter draws together some of the common themes of the book, including the adequacy of the law to cope with new forms of crime, the limits of the criminal process to deal with electronic theft, and the alternative or complementary countermeasures that may be relied on to maximize the benefits of electronic commerce, while minimizing the risks. As is the case with many domains of public policy nowadays, there are some issues that may be more amenable to resolution, in part at least, by market forces. A central issue of the conclusion will be the limits of the law in controlling electronic theft.

The focus of this book is generally on English-speaking democracies in the industrialized world, primarily Australia, the United States, Canada and the United Kingdom. Given the nature of the technology, the concerns of these nations are already shared more widely, and are likely to become even more common to other participants in the global economy in the years ahead.

Note

1 See 'Internet ban to protect Jews: Orthodox rabbis believe the Internet is a Trojan Horse for secular filth' (*Sydney Morning Herald*, 2000).
On the other hand, one could argue that digital technology may well contribute to the revitalization of small and vulnerable cultures. Today, Indigenous communities around the world have their own websites. See, for example,
http://www.cwis.org/wwwvl/indig-vl.html (visited 24 November 2000)
http://www.tpk.govt.nz/links.htm (visited 24 November 2000)
http://www.applicom.com/confeniae/english/ (visited 24 November 2000)
http://abc.net.au/message/links.htm (visited 24 November 2000)
http://www.inusiaat.com/icc.htm (visited 24 November 2000)
http://www.nativeweb.org/resources/organizations/native_american/ (visited 24 November 2000).

CHAPTER 2

Stealing Funds Electronically

Theft, in its simplest and most traditional form, is the dishonest taking of another person's property without their consent and with the intention of permanently depriving that person of the property taken. In the past, money, in the form of banknotes and negotiable instruments such as cheques, was often the primary target of thieves and robbers. The techniques used by acquisitive offenders were often simple and direct – extending from pick-pocketing, through bank robbery, to counterfeiting of cheques and currency.

Since the development of computers, however, theft of money can be carried out in a much wider range of ways, most of which can be undertaken entirely electronically by diverting funds from one bank account to another by means of the dishonest manipulation of funds transfer systems. We need to be clear, however, about what is involved in the electronic transfer of funds. Bags of money do not travel in dematerialized form along telephone wires (see Mackenzie 1998: 22). Instead, a series of instructions are given to financial institutions to debit and credit accounts of the parties to the transaction. Systems are then put in place to ensure that these instructions cannot be altered without the express authority of the senders and that other instructions cannot be substituted unlawfully. Fraud prevention simply requires that the instructions given by the parties to a transaction, be they consumers, merchants, or financial institutions, cannot be tampered with.

This chapter considers the problem of funds that are stolen electronically – the clearest and arguably most lucrative form of unlawful acquisition in cyberspace. As we will see in Chapter 4, millions of dollars have been stolen from government departments by public servants in recent years simply by diverting payments from legitimate to illegitimate bank accounts. The future will see opportunities of this nature extend dramatically, within both the public and private sectors. The many other problems arising from electronic commerce, such as those relating to misleading and deceptive practices that create particular concerns for consumer protection agencies, are dealt with in Chapter 7. Instead, the present chapter focuses on the ways in which the various electronic payment systems may be abused dishonestly for personal gain.

Electronic Commerce

The extent to which individuals and corporations are using computing and communications technologies for business transactions is increasing enormously. As we have seen in Chapter 1, the market is potentially great, with millions of users worldwide transacting business estimated to be worth trillions of dollars. Although the rate of increase in business-to-consumer electronic commerce has slowed somewhat recently, it will still represent an important part of economic life in the years to come. Business-to-business electronic commerce is likely to become the dominant means of transacting business in the short term.

Surveys of Internet usage have shown that, at present, most online consumer transactions involve small-value purchases such as books, CDs, wine, computers, and information technology products. The potential exists, however, for anything to be purchased electronically and we have recently seen the establishment of a number of online auction houses which deal in goods of much higher value (e.g. eBay 2000). A large proportion of Internet consumers also arrange travel and holidays electronically. Jupiter Communications has, for example, estimated that the online travel market will be worth US$16.6 billion by 2003 (Jupiter Communications 1999). In the future, transactions of much higher value will be conducted electronically, including the purchase of motor vehicles and house properties. This will make the theft of funds electronically even more attractive than at present.

Most business-to-consumer transactions on the Internet take place by purchasers identifying goods and services they require by inspecting Internet sites. They are then able to pay for the product chosen using conventional forms of payment such as money orders or cheques, which may be sent by post to the merchant before the product is despatched or the service provided.

Alternatively, purchasers can pay for a product or service by transferring funds electronically. This can be done by disclosing one's bank account (usually a credit card account) details and authorizing the merchant to debit the specified account to the value of the sum in question. In this way one could, for example, sitting in Melbourne, purchase a book from an antiquarian bookshop in London, or subscribe to an adult Internet site emanating from Copenhagen.

More recent payment systems have been developed that make use of telephone accounts to allow merchants to get access to a user's funds, as well as forms of electronic cash in which value is held electronically on the computer's hard drive and debited or credited as and when the need arises. New forms of stored-value cards, which have been designed to record monetary value, may also be used to transfer funds from a bank's ATM to a personal computer and thence to a merchant. These systems are obviously more efficient since transactions may be carried out and paid for instantaneously, allowing such activities as online gambling to take place.

The Risks of Electronic Commerce

Each of these payment systems, however, creates security vulnerabilities, and already we have seen instances of fraud being perpetrated electronically and funds being stolen online. In a telephone survey of 1006 online consumers conducted for the National Consumers League in the United States between April and May 1999, 24 per cent of those surveyed said that they had purchased goods and services online. Seven per cent, which represents six million people, however, said that they had experienced fraud or unauthorized use of credit card or personal information when conducting transactions (Louis Harris & Associates Inc. 1999). The risk of electronic theft is one of the principal deterrents to the development and expansion of electronic commerce. The Internet usage surveys carried out by the Australian Bureau of Statistics (1998, 1999, 2000) found that 5 551 000 adults (41 per cent of Australia's adult population) gained access to the Internet in the twelve months to November 1999, compared with 4 230 000 adults (32 per cent) in the twelve months to November 1998. In the twelve months to February 2000, approximately six million adults (43 per cent of Australian adults) gained access to the Internet (an increase of 11 per cent in the percentage of adults between November 1998 and February 2000).

The surveys also found that nearly 5 per cent of Australian adults (653 000) used the Internet to purchase or order goods or services for their own private use in the twelve months to November 1999. This represents a significant rise from the 2.6 per cent of adults (347 000) who did likewise in the equivalent period to November 1998. In the twelve months to February 2000, the percentage of adults who used the Internet to purchase or order goods or services for their own private use remained at 5 per cent (740 000 adults), an increase of 260 000 over the number of Internet shoppers in the twelve months to February 1999.

The number of people who paid for goods and services by giving credit card details online, however, dropped from 80.5 per cent (279 000 Internet shoppers) in the twelve months to November 1998 to 77 per cent (502 810 Internet shoppers) in the twelve months to November 1999. In the twelve months to February 2000, 74 per cent (547 600 Internet shoppers) paid for all or part of their purchase by giving credit card details online. This reduction is perhaps indicative of the concern that exists in the community regarding the security of online payment mechanisms.

The security risks associated with conducting online transactions may best be described by considering separately each of the three payment systems that are available: paper-based payment systems; direct debit electronic funds transfer (EFT) systems; and systems that make use of electronic cash.

Although some of these systems have yet to commence large-scale operation, security flaws have already been identified by which offenders are able to steal funds electronically. Although the following discussion describes the nature of the risks in general terms, it would be inappropriate to indicate precisely how some fraudulent activities are perpetrated. The present book seeks to inform the discussion of these issues constructively, and not to assist potential offenders in carrying out their activities.

Paper-based payment systems

Where goods and services are obtained through the Internet and paid for using paper-based instruments such as postal orders or cheques, fraud may be perpetrated in the same ways as those that have operated in the past where these payment systems have been used. Here the weak spots lie mainly in individuals using accounts that have been opened by giving false identification details, exceeding the credit balance held in cheque accounts, or counterfeiting or altering instruments themselves. Because there is pressure for transactions to be carried out quickly on the Internet, merchants may be less willing to wait for cheques to be cleared or for authentication checks to be carried out before authorizing the despatch of goods or the provision of services, thus leaving them open to fraud. Similarly, consumers may send off a cheque to a merchant located in a foreign country, about whom they have no independent information, who may receive payment and default on the agreement.

Direct debit electronic funds transfer systems

In addition to paper-based transactions, online payments could be made by direct debit, in which value is transferred directly from the payer's account to the recipient's bank, or by way of credit transfer in which a payer advises her bank to debit her account with a sum that is electronically credited to another account. These are essentially 'card not present' transactions which operate the same way as any telephone or mail-order transaction based on a credit card account.

In order for such transfers to take place, preliminary steps need to be taken by the parties involved, which include the exchange of account details and the conduct of various identification checks. From the purchaser's point of view, an element of risk arises if funds are transferred before the goods arrive or the service is provided. From the merchant's point of view, it is necessary for funds to arrive before the goods are despatched or the service provided. Both purchasers and merchants may also incur bank and government fees in respect of such transactions; these are higher in the case of Internet transactions, where the risks are greater.

The principal safeguard against such fraud is for merchants to take adequate steps to authenticate the card details and the card-holder, and to ensure that adequate funds are held in the account to cover the purchase. Getting authorization from the card issuer is the first step in preventing fraud, and some banks are now offering real-time authorization for transactions above the specified floor limits. Merchants should also require customers to sign receipts for any goods delivered, though this is impractical with some online transactions, for example the purchase of software, which can be downloaded immediately.

Home banking

The Internet is also being used for home banking, in which various transactions can be carried out from a personal computer in the customer's home which is connected to the bank via telephone wires and a modem. Home banking services include obtaining general information such as locations of branches and ATMs, interest and exchange rates; conducting various transaction services such as getting account balances and details of past transactions, transferring funds between accounts, using electronic chequebooks to pay bills; and other services such as ordering statements and chequebooks, reporting lost or stolen cards, notifying changes of address, stopping payment of cheques, obtaining loans, investment information or share dealing brokerage, and seeking share portfolio management services.

In each of these systems, protection against fraud is obtained through authentication based on a PIN, transaction codes, and encryption (scrambling) of data in much the same way as an ATM system operates. The possibility exists, however, that passwords, PINs and cryptographic keys could be compromised. To limit the financial consequences of this, banks issue a PIN for home banking that is different from the customer's ordinary ATM PIN.

Both telephone and computer home banking are now being offered in a number of countries, and already police have been called on to investigate allegations of fraud. In one case in Australia, two individuals claimed that they could illegally get access to Advance Bank's Internet banking system. They offered to solve flaws that allegedly existed in the bank's security system in return for a payment of A$2 million, thus amounting to a form of extortion. No prosecution occurred, however (Da Silva 1996).

Electronic funds transfer systems

Various systems are being developed to enable customers, banks and merchants to communicate securely with each other. A number of electronic funds transfer systems already operate throughout the world as substitutes for paper-based cheque transactions, and these could well be adapted for Internet use. The United Kingdom GIRO system, for example, has benefits in preventing cheque fraud because the payment order is directed to the banker directly rather than through the payee. In the GIRO system, the person wishing to make a payment, the payer, instructs his bank on the details of the payment and the funds are electronically transferred from the payer's account to the payee's account. These systems create a security risk if procedures are not in place to verify the availability of the funds that are to be transferred or if account access controls are not in place. There is also the possibility of information being manipulated as it passes over the network in unencrypted form.

In order to secure electronic funds transfers, data are generally encrypted using algorithms which encode messages. These are then decoded using electronic keys known to the sender and the recipient. The main security risk here lies in the possibility of the encryption keys being ascertained, in which case data within the system could be revealed or manipulated. Most of the large-scale electronic funds transfer frauds that have been committed in the past have involved the interception or alteration of electronic data messages transmitted from the computers of financial institutions (Meijboom 1988). In many cases offenders have worked within financial institutions themselves and been privy to the operation of the security systems in question. See, for example, the cases of Stanley Mark Rifkin (Rawitch 1979 and Sullivan 1987); *R* v *Thompson* [1984] 1 WLR 962; *Director of Public Prosecutions* v *Murdoch* [1993] 1 VR 406; and the 1994 Citibank case *R* v *Governor of Brixton Prison; Ex parte Levin* [1996] 3 WLR 657. See also Holland (1995) and Kennedy (1996).

In order to enhance the security of online transactions, various companies have designed systems to ensure that the identity of the contracting parties can be authenticated and that merchants can ascertain if the customer has adequate funds with which to conduct the transaction. The Secure Electronic Transaction protocol (SET), for example, uses public key encryption to protect data from being compromised. Digital signatures are also used to authenticate each of the parties involved. Credit card details are encrypted before transmission, with the decryption keys being separately protected. Merchants receive payment by passing to their bank an encrypted message that originates with the card-holder, allowing funds to be transferred from the credit card account to the merchant's account. In addition, card-holders are able to validate that the Internet merchant is legitimate through the merchant's digital certificate. SET software automatically checks that the merchant has a valid certificate thus confirming their relationship with their financial institution (Visa International 2000).

The main security risks associated with these systems relate to the possibility that private encryption keys could be stolen or used without authorization by people who have obtained them illegitimately. The easiest way to do this would be to submit false identification evidence to registration authorities when obtaining a public-private key pair. Alternatively, if a private key were held on a smartcard it might be possible to obtain access to the key simply by breaking the access control device on the card, which could simply be a password. Thus it could be possible for someone to make use of another person's private key to order goods or services from the Internet and be untraceable.

Card-based systems

Clearly, it would greatly facilitate electronic commerce if a user were able to insert a plastic card into an EFTPOS terminal attached to a personal computer and to conduct transactions directly between a merchant and a financial institution. This would, however, require that every personal computer be included in the computer network that links all financial institutions worldwide. Even if this were financially possible, plastic card payment systems have their own vulnerabilities to fraud through counterfeiting, alteration and theft of cards (Smith 1997), not to mention the logistic and security problems associated with having every financial institution's secure network provided to every Internet user.

Others are considering the use of smartcards that have the capacity to store value and transfer this to merchants via the Internet. Smartcard payment systems may take a variety of forms. The system that most closely resembles the early forms of stored-value cards involves a scheme operator which administers a central pool of funds. When a card-holder transfers value to the card, the funds are actually transferred to a pool controlled by the scheme operator. A merchant who is paid from the card takes evidence of the receipt to the scheme operator, which then pays the relevant amount from the pool. Other proposals, such as those operated by MasterCard and Visa International, envisage a number of brands of cards being accepted. In such schemes there is no central pool of funds, but rather each card issuer is responsible for reimbursing merchants who accept their cards.

In the United Kingdom, the Mondex system developed by the National Westminster and Midland Banks does not involve scheme operators. Funds are loaded onto the card, which can then be used without reference to any other person. Funds are transferred from one card to another as well as to merchants, but because funds loaded onto the card do not exist anywhere other than on the card, there is no audit trail of transactions or reconciliation of payments. This means that forgery could occur without trace and the scheme could be used for money laundering, or dispersing the proceeds of criminal activities. The Mondex card can be recharged from a mobile telephone link and can be used in EFTPOS terminals. In the United States, a modified version of the Mondex system is being trialled that will enable banks to trace card use. It will also be possible for money held on cards to be downloaded into computers, thus enabling Internet purchases to be paid for electronically from the card (Hansell 1996).

The main security risk associated with smartcards is the way in which data are encrypted. Levy (1994), for example, describes the potential for fraud using Mondex smartcards in England as being related to the cards' code encryption system. The encryption used on smartcards is able to be broken if certain types of errors can be created on the card, such as through the use of ionizing or microwave radiation. Bellcore, a United States computer and communications security company, and others have identified a number of design flaws in computer chip cards that may allow data to be leaked or information contained in the card to be tampered with (Bellcore 1996; Spinks 1996; Denning 1999).

Electronic cash

Various systems are also being developed that will permit transactions to be carried out securely on the Internet through the use of electronic cash, or value tokens which are recorded digitally on computers. The Digicash system, for example, which is based in the Netherlands, uses a form of electronic money known as 'E-cash' (Digicash 2000). Before purchases can be made, both the merchant and the customer need to establish banking arrangements and Internet links with the bank issuing the E-cash. The customer first requests a transfer of funds from his bank account into the E-cash system; this is similar to withdrawing cash from an ATM. The E-cash system then generates and validates E-cash coins which the customer is able to use on the Internet. The coins are data streams digitally signed by the issuing bank using its private key. The customer is then able to send E-cash to any merchant who will accept this form of payment using the software provided by the E-cash service provider. The customer encrypts the message and endorses the coins using the merchant's public key. The merchant then decrypts the message with its private key and verifies the validity of the coin using the issuing bank's public key. The merchant is then able to turn E-cash into real funds by presenting the E-cash to the issuing bank with a request for an equivalent amount of real funds to be credited to the merchant's bank account.

Another approach is being used by an Australian company, Cybank (1997), which has patented a scheme in which value is created online through the use of telephone accounts. By selecting a cash amount on the Cybank Internet site, the software dials a pay service number and the sum requested is debited to the customer's telephone account. The customer is then able to use the credit to purchase goods and services over the Internet (Bowes 1996; Carter 1996).

Preventing the Electronic Theft of Funds

Preventing funds from being stolen electronically will involve the use of both conventional legal responses and novel technological approaches. Although the possibility of criminal prosecution and punishment should always be present, business solutions founded in practical fraud prevention policies are likely to represent a more effective solution to electronic funds transfer crime. The importance of a pluralistic response to the theft of funds electronically cannot be over-emphasized. The legal reforms required for the implementation of, for example, a public key infrastructure that makes use of digital signatures must be accompanied by appropriate industry codes of practice and user education. Together, it should be possible to minimize the risk of large-scale abuse of electronic funds transfer systems.

Risk awareness and user education

One of the most effective strategies used to control crime is education of the public about the nature of the security risks they face, and how they can best protect themselves. As most online payment systems require the use of a PIN or password to gain access to personal computers or plastic cards, protection of access codes is the main crime prevention strategy available. Users are best placed to protect themselves by taking basic security precautions to ensure that cards are not stolen or their security features compromised. Consumers are advised against disclosing access codes, keeping them with cards or writing them on cards, and severe financial consequences result from failure to heed this advice. Studies reveal, however, that between 20 and 70 per cent of people write their PIN on the card or on a piece of paper carried with the card (Sullivan 1987).

Publicity and education campaigns have, however, been effective in reducing economic crime. In the United Kingdom, for example, a campaign by the Association for Payment Clearing Service resulted in a 41 per cent reduction in plastic card fraud between 1991 and 1994, while losses occurring at retail points of sale were reduced by 49 per cent during the same period. Losses from cards lost or stolen in the post were reduced by 62 per cent between 1991 and 1994 (Webb 1996). Had these fraud prevention initiatives not been introduced, it has been estimated that losses in Britain would have been 350 per cent higher than those recorded in 1996 (see Levi and Handley 1998).

When funds transfer systems become fully operational on the Internet, there will also be a need for users to be educated in ways in which they may protect themselves from fraud and deception. The Internet itself may prove to be the most effective mechanism for transmitting such information.

Institutional practices

Financial institutions and institutions involved in maintaining the infrastructure of electronic commerce are able to adopt a wide variety of self-help strategies that may reduce the risk of funds being stolen electronically. First, and most importantly, is the need for organizations to be confident that the staff they are employing are reliable and trustworthy, as electronic fraud often involves confederates with inside knowledge of an institution's security and computer procedures. Personnel should also be regularly monitored, particularly long-term employees who have acquired considerable knowledge of an organization's security procedures. Caution is also needed when organizational disputes develop.

One Australian case involved a 27-year-old male, known as 'Optik Surfer', who was sentenced to three years' imprisonment (with eighteen months suspended) on 27 March 1998 in Sydney for eight counts of obtaining unlawful access to a computer and one count of unlawfully inserting data into a computer. The offender, who was a computer networking consultant, had been refused employment with an ISP in January 1994, and in March 1994 took revenge by illegally obtaining access to the company's computer network using the user account and password of the company's technical director. He then gained access to the company's database of 1225 subscribers and publicized their credit card account details by disclosing them to various journalists. He also altered the company's home page on 17 April 1994 by including a message that the company's security system had been compromised. The publicity resulted in the company losing more than A$2 million through lost clients and contracts. It had to change its business name and sold the Internet access part of its business to another ISP (*R* v *Stevens*, unreported decision of the NSW District Court, 27 March 1998).

Financial institutions may be able to assist online merchants by notifying them of incidents of fraud as soon as they occur, so that they can avoid repeat victimization or so that others can avoid victimization. Systems may also need to be created by which online merchants are able to obtain immediate authorization from financial institutions before transactions are accepted. It may even be necessary for all online transactions to be authorized before they are accepted.

Frauds in which merchants are involved constitute a large problem for financial institutions because merchants or their employees are ideally placed to permit access to computer networks and to alter transaction details. Financial institutions may need to make use of artificial neural networks in order to isolate fraudulent claiming patterns by merchants and maintain databases of merchants who have engaged in illegal conduct on the Internet in the past. Already, organizations are providing certification services to enable users to identify illegal or unsafe Internet sites, as we shall see in Chapter 7.

Technological solutions

A wide range of technological solutions have been devised in order to reduce the security risks associated with conducting business electronically.

Hardware security

In order to provide a safe system for electronic commerce, computer hardware needs to be adequately secured. This extends from computer terminals used in homes, businesses, and public kiosks, through servers operated by ISPs, to the hardware maintained by merchants and financial institutions. The extent of the security precautions used will be determined by the risks present. Terminals located in Internet kiosks may need only basic access controls such as the use of passwords or smartcard tokens, while servers maintained by banks might need to be shielded against electromagnetic radiation (EMR) scanning.

The threat of EMR scanning should not be taken lightly. Although the risk is remote, the possibility exists. In one case in England, for example, a computer eavesdropper scanned electronic transaction information transmitted by a bank. Despite the fact that the information was encrypted, the code was defeated and the individual successfully obtained £350 000 by blackmailing the bank and several customers, threatening to reveal certain information to the Inland Revenue (Nicholson 1989).

If payment systems based on digital signatures and encrypted data transmissions are used, then the need to protect computer cables from interception would not arise as any data would not travel in clear text. At present, however, a good deal of sensitive information travels across networks in unencrypted form, making it vulnerable to interception and subsequent disclosure. The adequacy of encryption as a security measure depends, of course, on the strength of the encryption system used and the determination of the attacker.

Card security

Plastic cards may be used in conjunction with online transactions in a variety of ways, as we have seen. They will mainly be used to store access devices such as cryptographic keys or other user authentication devices. They may also be used to store value in Mondex-type smartcard systems. The most sophisticated security features should be built into plastic cards to prevent counterfeiting, alteration or unauthorized access. Newton (1995) describes various crime prevention strategies that have been used to prevent the counterfeiting of plastic cards including the use of security printing, micro-printing, holograms, embossed characters, tamper-evident signature panels, magnetic stripes with improved card validation technologies, and indent printing. Smartcards, of course, are much more difficult to copy than ordinary magnetic stripe cards.

Unfortunately, all of these card security features, including computer chip circuitry in smartcards, have been overcome by organized criminals. Online payment systems that do not rely on plastic cards will presumably be much more secure and it may also be possible for these to operate in conjunction with biometric user identification systems.

Value restrictions

As an alternative to target hardening (i.e. making it more difficult to act illegally by introducing various changes to the physical environment [Clarke 1995]), it has been suggested that the risk of large-scale fraud and money laundering using Internet-based funds transfer systems could be restricted by placing limits on the size of transactions.

Mackrell (1996), for example, has suggested that stored-value cards should have a modest limit placed on the maximum value that can be stored on them, especially if they are to be used for card-to-card transfers. There could also be a limit on the life of the cards which would restrict their usefulness for hoarding and money laundering. Self-expiring cards have also been developed that automatically deteriorate after a certain time. In the case of online commerce, electronic restrictions could be placed on the value of transactions in order to avoid the possibility of large-scale fraud, though this may be seen as an unwarranted intrusion into freedom of electronic commerce.

Password protection

Passwords, used as a means of authenticating users of computer networks, are popular at present but are often misused or abused. It is possible to guess passwords, particularly if little thought has been given to their selection, or to use various forms of social engineering to trick users into revealing their passwords for subsequent improper use.

The use of brute computing force has also been used to break passwords. Password-cracking programs are available by which computers are able systematically to search entire dictionaries in search of a password. Even if passwords are encrypted to protect them from direct exposure, encryption keys have been broken through the use of massive computing resources (Denning 1998).

There are various ways of enhancing access security through the use of passwords (see Alexander 1995). Appropriate education of users is a first step in which information is given on ways of ensuring that passwords are not disclosed, guessed, or otherwise compromised. Systems should be used that change passwords regularly, or that deny access after a specified number of consecutive tries using invalid passwords. Terminals should have automatic shutdown facilities when they have not been used for specified periods, such as five minutes. Single-use passwords, where the password changes with every successive login according to an agreed protocol known to the user and system operator, could also be used.

Challenge-response protocols may also be used as a means of carrying out user authentication. The server generates a random number which is sent to the card. In a public key system, the card digitally signs the number and returns it to the server; the server then validates the digital signature. Alternatively, call-back devices may be used. After the user dials into a computer through a modem and gives his identity, the system disconnects the user and then telephones him on a number previously registered with the server; after the user is verified, the transaction can proceed. Such a system can be overcome, however, through the use of call-forwarding arrangements (Denning 1998: 45).

Another user authentication system makes use of space geodetic methods to authenticate the physical locations of users, network nodes, and documents. One company, CyberLocator, operates a location signature sensor that uses signals transmitted by satellite to provide a location on earth at any given time. Users can thus be located at the time they attempt to gain access to the system, which provides a safeguard against individuals located in a different physical location pretending to be legitimate users (Denning 1998).

Biometrics

One way in which problems of password and token security may be overcome is for users to identify themselves biometrically. Already there is a wide variety of such systems being used, all of which make use of a person's unique physical properties. Common biometric identifiers today include fingerprints, voice patterns, typing patterns, retinal images, facial or hand geometry, and even the identification of a person's subcutaneous vein structures or body odours (Johnson 1996).

Fingerprint identification systems are now being used in retail stores and for access to ATMs (Anon. 1996), while in California, a company called Identix has developed a system which has fingerprint recognition sensors on mobile telephones, computer keyboards, and plastic cards (Young 1999). The Bank of Texas has also introduced iris recognition systems for its ATM network.

The costs and volume of data required to be stored online to enable comparison for any potential user may, however, be prohibitive and there is always the possibility that computer security systems could be compromised by reproducing data streams that correspond with the biometric characteristics in question. Another problem is that users are required to provide samples of their personal characteristics but that the security of these samples could be compromised. Recognition systems are also at present – though not presumably in the future – costly and sometimes slow to use.

Digital signature security

The use of public key encryption systems also have their security risks. Public key systems require that cryptographic key pairs be issued to individuals who can establish their identity well enough by supplying several independent sources of identification such as those required when bank accounts are opened. Primary documentation (such as a passport, birth certificate, etc.) along with matching secondary documentation (such as a bank statement, car registration papers, etc.) would satisfy the required degree of documentary evidence of identity. But this may prove one of the system's weakest points. Already systems that require the identification of individuals when they open accounts with financial institutions have been circumvented by offenders producing documents that have been forged or altered through the use of computerized desktop publishing equipment.

There are various solutions to the problem of counterfeit identification documentation. First, and perhaps most important, is the need to validate identification documents with the issuing source. Staff presented with a birth certificate should, for example, check if the details correspond with those held in the central office of Births Deaths and Marriages. An electricity account tendered as an identification document should be validated by checking with the electricity company concerned. This may not always solve the problem, however, as telephone answering services can be manipulated to support the creation of false employment or identity details.

Second, staff who validate documents need to be instructed on the security features of documents, what original documents look like, and how they are forged and what forged documents look like.

Third, modern security features should, if at all possible, be incorporated on all primary and even secondary identification documents. Among these new technologies are security printing, in which colour-coded particles are embedded into the medium, 'tracer fibre' which can be woven into textile labels, and hidden holographic images which can be read with a hand-held laser viewer or machine reader, thus permitting verification of a product's origin and authenticity. Similarly, technologies have already been developed that make counterfeiting extremely difficult.

In any system that entails the identification of individuals to whom cryptographic keys are to be issued, these primary fraud prevention procedures would need to be included. Once questions of identification have been resolved, issues would arise as to the manner in which keys or hardware tokens are given to users. Standards would also need to be complied with for the storage and use of keys, perhaps by requiring keys to be used offline or with a smartcard that is able to process transactions.

The problem remains, however, that private key data or tokens themselves must be communicated to users. The financial world has already experienced considerable problems in transferring possession of plastic payment cards to users, and similar problems could arise with cryptographic keys that are stored on smartcards. There would need to be adequate security precautions to ensure that tokens are passed securely to users from the issuing authority.

Another area of risk is the generation of cryptographic keys. It may be possible for a person who generates a public and private key pair to retain a copy of the private key for later illegal use. Legislation may be needed which will hold the key generator liable for subsequent losses that arise out of the compromise of a key issued by that generator. Cryptographic keys would be kept on the hard drive of a computer, with the cryptographic service activated by a smartcard inserted into the PC. Smartcards may also be used to sign a digital signature and to authenticate the identity of a user.

In addition to the risks associated with compromising access mechanisms such as PINs, passwords and biometric devices, it is possible that smartcard tokens themselves may be altered or counterfeited. This has already happened with smartcards used for small-value commercial transactions. Where keys are stored on personal computers or servers, their security may be compromised, in which case appropriate risk management measures need to be taken.

Fraud detection software (neural networks)

If online fraud cannot be entirely prevented, it may at least be possible to identify fraudulent transactions quickly in order to reduce any losses suffered or prevent a repetition of the incident. A number of organizations are now providing software for the prevention of electronic funds transfer fraud.

Software has been devised that analyses user spending patterns in order to alert people to unauthorized transactions; there are also merchant deposit-monitoring techniques to detect the claiming patterns of corrupt merchants. The success of such an approach, however, depends on the extent to which the software cannot be interfered with or modified.

Legal Responses

As already mentioned, traditional methods of fraud prevention and the technologies of target hardening can succeed only to a certain degree. Commercial and business solutions need to be supported by legal responses, both of the hard (legislative) and softer (codes of practice) kind.

Systems for the electronic transfer of funds have, however, created a number of problems for the effective prosecution of offenders that uses conventional property laws. The manipulation of computers that control ATMs to enable cash to be dispensed when accounts held insufficient funds, for example, created legal difficulties in the 1980s in Australia, some of which led to the alleged perpetrators escaping punishment (Grabosky and Smith 1998: 160–7). More recently, in the United Kingdom, the Theft Act 1968 had to be amended in 1996 by adding section 15A to ensure that the fraudulent transfer of funds from one bank account to another could be prosecuted effectively. The problem arose following the decision of the House of Lords in *R* v *Preddy* ([1996] AC 815) that the electronic transfer of funds from a building society's account to the account of the defendant did not involve theft of the funds since 'property belonging to another' had not been stolen. The House of Lords found that the defendant in the case had not obtained the lending institution's chose in action (debt), but rather the electronic transfer of funds to the defendant's account had created a new chose in action which was a debt owed by a different bank to the defendant (see Chapman 2000).

The development of secure systems for electronic commerce have also required some property laws to be rewritten. A range of different approaches has been taken to law reform internationally in order to accommodate electronic commerce, with some parliaments enacting highly specific reforms that define 'documents', 'writing', and 'signatures' as well as specifying the rules that govern the attribution of communications. In Australia, a more general approach has been adopted with the enactment of the *Electronic Transactions Act* 1999 (Cwlth), which provides broad, technology-neutral provisions that constitute a basis for more specific legislative changes to follow. Even legislation written in technology-neutral terms is likely to warrant amendment to deal with new systems of electronic commerce – such is the rapidly changing architecture of digital technology in this field.

Despite the legal difficulties of mounting a criminal prosecution in cases in which funds have been stolen electronically, offenders continue to be apprehended, prosecuted and convicted, sometimes receiving sentences of imprisonment or substantial fines. In Australia, for example, an employee of a large corporation, who had illegally gained access to the corporation's computer network and altered billing records so as to perpetrate a fraud, was convicted in 1998 and sentenced to two and a half years' imprisonment (Australian Federal Police 1998: 30).

In the United Kingdom, the largest bank fraud ever committed, involving approximately £800 million, resulted in one offender being sentenced to fourteen years' imprisonment on 8 May 1997. The offender was found guilty on 3 April 1997 of fraud and false accounting after a massive investigation by the Serious Fraud Office and the City of London Police. Hundreds of thousands of people around the world lost their savings when the Bank of Credit and Commerce International subsequently collapsed. The offender was also sentenced to a further three years' imprisonment if he failed to pay a £2.94 million compensation order within two years. Although he was legally aided, he was ordered to pay prosecution costs of £4.3 million (Serious Fraud Office 1997).

In another case carried out between December 1995 and May 1996, thirteen men in England perpetrated a nationwide scheme to steal money from the banking system by manipulating the Clearing House Automated Payment System (CHAPS). The offenders, under the pretence of wanting to buy a product from the target company, asked over the telephone for the company's bank details, saying they intended to pay by CHAPS. They then forged the company's letterhead and directors' signatures and directed the company's bank to transfer money out of the company's account into the bank accounts of accomplices. The accomplices were willing to have their bank accounts used in this way for a percentage of the takings. The fraud involved 131 attempts to obtain funds, 33 of which were successful, totalling £1 906 748. During 1997, at Birmingham Crown Court, thirteen of the fourteen defendants either pleaded guilty to or were convicted of a number of offences such as conspiracy and obtaining property by deception. They received sentences ranging from three years' imprisonment to a £1000 fine. A BMW car that was used in the fraud was also confiscated, as was £32 000 worth of assets belonging to one defendant (Serious Fraud Office 1998).

The extent to which lengthy prison terms constitute a deterrent to fraudulent conduct is, however, open to debate. While many property offenders behave more or less impulsively, white-collar offenders are more likely to engage in rational calculation, making some assessment of the prospective benefits and costs of any fraudulent course of action. In these circumstances, the greater the perceived likelihood of conviction and the more severe the expected punishment, the less the inclination to offend. People who are aware, for example, that their assets may be confiscated after a criminal conviction may consider that the benefits to be derived from offending are not worthwhile. Arguably, the continued use of assets forfeiture legislation such as that which operates under Australia's *Proceeds of Crime Act* 1987 (Cwlth) is beneficial and deserves increased publicity.

It is evident from the extensive research that the possibility of imprisonment acts as a general deterrent, though to an unsubstantiated extent, for some members of the community, and that well-publicized severe sentences imposed on the most serious white-collar offenders help send the important message to the community that dishonesty is not tolerated. Imprisonment, however, is not a panacea. Increasing the number and length of sentences will not necessarily achieve commensurate decreases in crime. Indeed, for some individual offenders, disqualification as a company director may be a far more effective sanction than a severe fine or even a short term of imprisonment. Adverse publicity can also have profound effects in terms of shaming an offender in the community, perhaps more so than undertaking anonymous community service.

Apart from legal problems, there are a number of impediments to mobilization of the law. Because fraud offences perpetrated electronically do not involve face-to-face contact, it is possible for offenders and victims to be located in different jurisdictions. A number of sophisticated conspiracies have involved individuals in three or more jurisdictions, creating considerable problems for those investigating offences. Even if evidence of an offence can be adduced, the chance of locating the fraudsters, obtaining extradition and conducting a trial may be both costly and difficult.

Moreover, even where a perpetrator has been identified, two problems arise that have an international aspect: first, the determination of where the offence occurred in order to decide which law to apply; and, second, obtaining evidence and ensuring that the offender can be located and tried before a court. Both these questions raise complex legal problems of jurisdiction and extradition (see, generally, Lanham 1997). The solutions to these problems require the international harmonization of substantive laws and laws of evidence as well as the creation of mutual assistance arrangements to enable evidence to be taken abroad.

In Australia, the problem of harmonizing laws relating to electronic commerce and the prosecution of dishonesty offences is being addressed by the Commonwealth Attorney-General's Department, which is preparing a uniform Model Criminal Code for implementation throughout Australia. Once enacted throughout the country, the provisions of the code should ensure that electronic fraud offences are able to be prosecuted regardless of the technologies used in their commission or the geographical location of the offenders, provided that there is some connection with the jurisdiction in question (see Commonwealth of Australia, Model Criminal Code Officers Committee of the Standing Committee of Attorneys-General 2000).

As an alternative to legislative regulatory controls, the banking and credit industries have relied on codes of conduct to prevent fraud and to resolve disputes between institutions and customers. Codes have the dual function of acting as a form of education and publicity for institutions and customers, as well as providing a statement of recommended practice that can be relied on to resolve individual disputes.

Codes of conduct are, however, only going to be an appropriate regulatory mechanism where financial institutions or system operators are involved. Where electronic money or stored-value cards are used on the Internet, only the consumer and the merchant may be involved, and for these, existing codes of conduct may have to be rewritten.

It may now be appropriate for an Internet Commercial Code of Conduct to be established to deal with the allocation of risk and the determination of liability in Internet-based transactions. Issues that could be dealt with in such a code could include: guidelines on users' obligations in maintaining computer hardware in a secure environment; principles to be observed for obtaining, storing and using encryption keys securely; principles to be observed for storing key tokens securely and for preventing unauthorized access to them; obligations to be complied with regarding security and privacy of data; and principles to be observed in determining liability and the allocation of loss arising out of the use of the Internet.

Failure to comply with the provisions of a code of conduct could result in powerful economic and business sanctions being meted out to non-compliant entities and could also be relevant for establishing breach of civil legal obligations such as duty of care, insurance, or contractual compliance. In the future, some criminal consequences could follow repeated or serious failure to comply with the provisions of a code of conduct.

Conclusion

This chapter has described various ways in which funds may be stolen by exploiting security flaws in electronic funds transfer systems. The range of electronic systems used to conduct commercial transactions is increasing rapidly and considerable effort is being directed at ensuring the security of digital transmissions which represent monetary value. The opportunities for fraud are, however, substantial.

The solution to this kind of crime will ultimately mean the adoption of strategies in which close cooperation will exist between all those involved in providing and using systems. This includes telecommunications carriers and service providers, financial institutions, retail merchants, and individual users.

One area of particular importance is the need for strategies that will enable the weaknesses in systems to be quickly identified. Once recognized, there should be a prompt response to the problem. In ensuring that particular weak points in security systems are identified and weaknesses solved, it is likely that technology will provide the most effective means.

Probably the greatest area of risk in conducting online business is user authentication. Identity-related fraud has been a continuing problem in commerce for decades now and it is likely that it will continue in adapted forms on the Internet. If passwords continue to be used to restrict access to computers, they should be protected by the various security devices mentioned. Biometric identifiers will presumably become much more widely accepted as electronic commerce develops.

In planning for the future, it will be necessary to ensure that the weak points in security protocols are not overlooked. As in other areas of fraud control, the weaknesses in electronic commerce will invariably arise from human rather than technological factors.

CHAPTER 3

Digital Extortion

The term 'extortion' means many things to many people. Some extreme libertarians use it to refer to taxation in general, perhaps a reflection on revenue-raising practices of years long past; thirteenth-century English law, for example, prohibited extortion by sheriffs and other royal officials (Lindgren 1983). Today, some use it to refer to extremely high prices or interest rates, whether or not these are legally permissible. Most recently, it has been used to refer to the practice of registering famous company names as Internet domain names, then offering to sell the rights to the company in question (Fenwick and West 1998). Such terminological looseness is unfortunate, and makes constructive analysis of the problem rather difficult.

Extortion Defined

Conceptually, extortion is a close cousin to robbery and accordingly entails an element of theft. While robbery entails theft accompanied by the use of force or the threat of imminent force, extortion refers to a demand accompanied by the threat of unpleasant consequences at some time in the future. Not all jurisdictions draw a clear line between robbery and extortion, and some complicate matters further by specifying the offence in other terms, referring instead to theft generally, to criminal coercion, or to 'demanding money with menaces'.

Extortion is also related to bribery. In contrast to a bribe, where payment is proffered voluntarily in return for a favourable consideration, extortion entails a demand for payment as a condition of favourable action (or as a condition of desisting from unfavourable action). For example, a police officer may offer not to arrest or charge an offender, or a prosecutor may offer to drop charges, in return for a payment. A very fine line exists between extortion, where the payment is coerced, and bribery, where the payment is voluntary. The basic issue here is the intent of the person making the payment; it is the element of coercion that differentiates extortion from bribery.

Also related to extortion is the practice colloquially known as 'loan sharking' (Reuter 1983: Ch. 4). Here the offender extends credit to the victim at interest rates usually in excess of those legally permitted. The element of extortion lies not in the interest rate charged but in the implied method of collection, which usually entails some form of intimidation, or worse, in the event of non-payment.

An extortion threat is not necessarily limited to bodily harm. It may extend to physical harm to a third party (such as a relative or friend), damage to property, or even the disclosure of information potentially incriminating or otherwise embarrassing to the victim. This latter form is often referred to as 'blackmail'. 'Extortion' and 'blackmail' are often used interchangeably, despite their different origins (originally, while both crimes involved gaining property by threat, extortion was committed by public officials and blackmail by private individuals). Some commentators seek to distinguish between the two, referring to extortion where the action threatened is independently illegal (such as the infliction of bodily harm), and to blackmail where the action threatened is independently legal (such as the disclosure of embarrassing information) (Wertheimer 1987: 91).

The threat of disclosure of information which the victim would rather be kept private is a common form of extortion. This could entail disclosure of previously undetected criminal activity, or otherwise embarrassing personal details. A familiar scenario involves the person who is threatened with disclosure of a previous or current clandestine sexual relationship unless he or she engages in specified conduct. In the absence of an existing secret relationship, enterprising extortionists may seek to fabricate one. The extortionist may thus lure the victim into a sexual or otherwise compromising relationship, then threaten him with disclosure.

In the case of the threatened disclosure of an unpleasant secret, there exists a paradoxical situation in which two elements of blackmail, taken independently, are legal, whereas taken jointly they are criminal. For example, one may lawfully threaten to report a businessman or woman's tax evasion to revenue authorities, and one may lawfully seek a lucrative contract from that businessperson. But packaged together, these constitute extortion. As one commentator has put it, 'The old saw has it that two wrongs cannot make a right. The paradox of blackmail is that two rights can make a wrong' (Wertheimer 1987: 91).

The Purpose of Extortion

As intimated above, the object of an extortion demand (in other words, the thing that is demanded) may or may not be financial, and may include a variety of non-financial considerations such as sexual favours, or favourable discretionary action of an official nature such as non-enforcement of certain laws. Extortion may also be an element in a wider project. Terrorist organizations may engage in extortion with various ends in view, demanding resources with which to support their operations, or seeking other concessions such as the release of persons in custody. They may also intend to destabilize the society or government. In the United States during the 1990s the Unabomber offered to desist from further bombing if his Manifesto were published widely (Sacramento Bee 2000).

The Targets of Extortion

Extortion threats may be directed at individuals or at organizations of various sizes. Where enrichment is the extortionist's objective, wealthy people and commercial enterprises are often selected as the target because of the significant assets at their disposal. But vulnerability is by no means limited to the rich. Small businesses may be at risk of extortion from entrepreneurial criminals. The most commonly cited examples are gangs who extort payments from small businessmen in Chinese communities (Chin 1996).

The nature of their position places public officials at risk of becoming either the victim or the perpetrator of extortion. Those officials who are in a position to dispense largesse may be inclined to do so only under conditions of reciprocity. Conversely, public officials whose past may contain embarrassing or incriminating secrets may be threatened with disclosure unless they deliver appropriate considerations.

The legal boundaries of extortion are not always distinct, and this has significant implications for investigation and prosecution. A public official charged with extortion may claim that she had been offered a bribe; a citizen charged with bribery may try to make a case that he had been threatened. As we shall soon see, the threat itself may be so veiled as to be implicit. In such circumstances the element of intent may be difficult to prove.

Conventional Manifestations of Extortion

The most familiar form of extortion in Western societies today is what is commonly termed 'protection'. Basically, the perpetrator offers to refrain from inflicting future harm on a victim in return for payment (Chin 1996). The perpetrator may seek partially to legitimize the arrangement by offering to protect the victim from future harm from a variety of other sources as well (Gambetta 1993). Although the extortionist may be part of a group, and the target may be an organization, extortion typically takes the form of a one-on-one transaction.

Since the element of threat is central to the charge of extortion, some perpetrators may try to structure a threat in such a way that it appears to be something else. Where the practice of offering 'protection' is routine, the threat may be implicit, and an extortionist may request a 'loan' or a 'donation' from the victim. Alternatively, extortion may be disguised within a commercial transaction, where a product, for example waste disposal or laundry services, is purchased at an exorbitant price – or else. In such instances the extortionist is making an offer that the victim can't refuse.

Protection may also be demanded of large organizations. Japanese criminal organizations often receive payment from large corporations to ensure that annual meetings of shareholders are not disrupted. Such organizations even have a name, sokaiya (Kaplan and Dubro 1986: 169).

Another common form of extortion is that which is integral to ransom kidnapping. Here the perpetrator will abduct a person and threaten him with harm unless the offender's demands are met. Closely related to ransom kidnapping are extortion plots involving threats to detonate a bomb (usually on an aircraft or some heavily populated public place). Yet another basic form of extortion relates to product tampering, where a particular consumer product is altered or damaged in such a way as to create a threat to public health and safety. The manufacturer of the product may be coerced into paying the extortionist, in order to protect commercial reputation.

Penalties for Extortion

The criminal laws of most jurisdictions treat extortion as a serious matter, with the degree of seriousness tending to vary with the circumstances of the offence. For example, demanding or receiving a consideration after a threat to inform against a violation of US federal law carries a fine and a prison term of up to one year (18 USC 41. s 873). In the Australian State of Victoria, a court official who extorts, demands, takes or accepts any unauthorized fee or reward can be fined and imprisoned for up to two years (*Magistrates' Court Act* 1989 (Vic.) s. 23). In the United States, interstate communications of a threat to kidnap or injure a person with intent to extort carries a fine and up to twenty years' imprisonment, while threats to injure property or reputation, a fine and a maximum of two years (18 USC 41 s 875 (b) (c) (d)). In Victoria, extortion involving a threat to kill or injure carries a maximum penalty of fifteen years' imprisonment, while an extortionate threat to destroy certain forms of tangible property (such as buildings and vehicles) carries a maximum of ten years (*Crimes Act* 1958 [Vic.] ss. 27–28).

Extortion in the Digital Age

Despite its long history, the crime of extortion has taken on some new faces in the digital age. The application of new technologies to extortion takes five basic forms. First, digital technology can be used as the medium to communicate an extortion threat. Second, given their extrinsic value, as well as a prospective victim's dependence on them, information systems themselves may be specified as the target of an extortion threat. Third, in circumstances where an extortion threat involves public disclosure of embarrassing or harmful information, digital technology, especially Internet-related technology, can be the medium through which this information is communicated. Fourth, when an extortion entails some financial consideration, electronic payment systems may be used to facilitate the payment of the proceeds of extortion or to conceal such payments. And finally, information technology may be incidental to the offence in question, such as when it may be used to gather information about a prospective victim. These various forms of digital extortion entail a range of legal consequences, some involving conventional property offences and others involving different kinds of illegality such as those that relate to computers and telecommunications. Our discussion considers the full range of conduct that might be generically described as digital extortion. We now turn to each of these five issues.

Information systems as the medium of threat

An 'offer that one cannot refuse' can be made in many ways, from verbal communication in a face-to-face encounter to the written word. The 'ideal' extortion threat, at least from the point of view of the perpetrator, is one that divulges neither the identity nor the location of the extortionist. Letters sent through the post will bear a postmark, or perhaps other distinguishing features, including fingerprints or DNA traces, which may help to reveal their source. Although stolen or cloned cellular phones can help conceal one's identity, there is a limit to which the 'anonymous' telephone call can remain anonymous, as telephone calls can often be traced. While in the past this required the reactive mobilization of tracing technology, the advent and increasing popularity of 'Caller ID' may enhance the transparency of telephonic communications, except the one-off use of stolen cell phones, or where the disclosure of caller information is suppressed, or the technology manipulated in some way to ensure anonymity.

Nevertheless, many Internet extortion attempts appear to show little effort to cover the extortionist's tracks. Ron Hornbaker, a 29-year-old veterinarian from Kansas, frequented an America OnLine (AOL) Internet chat room 'Married but Looking' and posed as a woman named Rita. He engaged in sexual banter with other users, then invited them to view erotic photos of Rita in private rooms. When the men entered the rooms they received threatening messages, supposedly from Rita's policeman husband. Hornbaker, posing as Rita's husband, threatened the men with bodily harm unless they paid an amount of money ranging between US$500 and US$2000. An Illinois man reported Hornbaker to the FBI; he pleaded guilty to extortion and was sentenced to two years in jail and a fine of US$1000 (Lash 1996; *U.S.* v *Hornbaker*, US District Court, Northern District of Illinois, 96 CR 50026, 16 August 1996).

The use of the Internet to communicate a threat may provide a modicum of anonymity, if sufficient precautions are taken by the extortionist. These might entail looping and weaving communications through various servers, and the use of anonymous remailers (Denning 1999: 218, 313–16). The global nature of cyberspace is such that Internet extortion threats may transit a number of nations *en route* from originator to target. But the fact that electronic communications are often traceable will increase the risk that the digital extortionist will be identified. The use of these techniques of anonymity and evasion may be compounded by other illegalities to impede investigation further. Denning (personal communication 1999) cites one case in which an extortionist threatened to fly a remote-controlled model airplane into a jet engine of an aircraft during takeoff from a German airport. The extortionist used an anonymous remailer in the United States, and accounts with America OnLine that had been established under an assumed name with fraudulent credit card details – somewhat like stealing a motor vehicle for use as a getaway car. Despite the elaborate efforts of the extortionist, the plot was foiled.

Beyond its use to communicate a threat, digital technology may be used *by the target* of an extortion attempt for the purpose of communicating *with* an extortionist. A recent example may be drawn from the 1997 extortion attempt against Microsoft chairman Bill Gates. A 21-year-old Illinois man, Adam Pletcher, sent Gates four letters through the regular mail of the US Postal Service demanding that a sum of money be deposited in an overseas bank account. He further directed that Gates reply initially through an Internet forum on America OnLine. The extortionist subsequently sent Gates a computer diskette with a request for additional information.

The letters were postmarked in northern Illinois. When investigators covertly emailed a reply to the AOL forum over Gates' name, they were able to identify those AOL subscribers from the region who were logged onto the forum at the time. In addition, analysis of the diskette revealed the names of two people who turned out to be the parents of the extortionist (Haines 1997). The suspect was charged and convicted in due course on four counts of sending mail with the intent to commit extortion, and sentenced to 70 months in prison, considerably less than the twenty-year maximum.

The elements of extortion may be combined, without being linked. The notorious hacker Justin Petersen, also known as 'Agent Steal', was convicted of having telephoned a bomb threat to a financial institution to ensure that wire transfer officers were away from their desks, then hacked into the system and paid himself US$150 000 (Associated Press 1998a).

Information systems as the targets of threatened action

It is now almost trite to suggest that humankind is becoming increasingly dependent on information systems. Whether they encompass such crucial infrastructure as electric power distribution, air traffic control systems, or the resources of a large financial institution, systems and the information stored in them are of immense value and are therefore potentially attractive targets for extortionists.

An important factor in any extortion attempt is the credibility of the threat (Ellsberg 1968). There exists, for example, a crudely phrased form letter, purportedly from an organization of highly trained assassins based in Nigeria, which threatens the recipient with execution unless US$35 000 is remitted to a Swiss bank account within 96 hours (419 Coalition 2000). Like many offers emanating from Nigeria, it is not terribly convincing. As is the case with the more common Nigerian advance fee frauds, the architects of the scheme no doubt expect a low success rate and base their strategy on mass postings.

Where threats are credible, the reputation of a company or of an entire industry may be put at risk. To an even greater extent than is the case with product tampering, an organization's public image can suffer dramatically because of electronic vulnerability. The trajectory of our progression to an information economy, and the public's confidence in the integrity of online sales or banking, can be jeopardized by the appearance of exposure to criminal exploitation. Denning (1999: 230) reports that many corporate web pages have been targeted by extortionists who threaten to post matter defamatory to public figures. To enhance the credibility of their threat, they make small changes to the company's page.

The London *Sunday Times* reported in 1996 that over 40 financial institutions in Britain and the United States had been attacked electronically over the previous three years. In England, financial institutions were reported to have paid significant amounts to sophisticated computer criminals who threatened to wipe out computer systems (*Sunday Times*, 2 June 1996). The article cited four incidents between 1993 and 1995 in which a total of £42.5 million was paid by senior executives of the organizations concerned, who were convinced of the extortionists' capacity to crash their computer systems (Denning 1999: 233–4).

In some extortion incidents, the element of threat may be suppressed, and the victim presented with a fait accompli, the resolution of which is conditional on some payment. One of the earliest reported cases of extortion targeting a computer system occurred in 1971 (Parker 1976: 20). A programmer was engaged to automate the systems of a company that provided accounting services. He completed the task and promptly disappeared, taking with him all copies of the programs and documentation. He then demanded US$100 000 in return for the materials. Obviously better at programming than at extortion, he was caught and charged within three days. Significantly, charges were later dropped when the prosecution saw it as too difficult to proceed with a matter as obscure as theft of computer programs.

This is the basic model for extortion by an inside employee. The essential form has not changed in the past three decades, though the capacity of courts to cope with intangible evidence may have improved somewhat.

The fait accompli may entail the insertion of some viruses, especially those so-called 'cryptoviruses' that encrypt data and make it inaccessible to the user. The traditional use of cryptography has been defensive in nature, to provide privacy and security; the offensive use of it can cause lack of access to information. Extortion based on cryptography could for example entail the offender's encryption of valuable data belonging to the victim, and offering to 'sell' the encryption key in return for a consideration (Young and Yung 1996).

In addition to the use of viruses or encryption, it has been suggested that information systems may also be targeted by radio frequency weapons. These are instruments which emit electromagnetic radiation that can interfere with or damage electronic equipment. Denning (1999) expressed scepticism that high-energy weapons have been used in those few cases of extortion that have reached public attention. It does appear, however, that the technology is, or will soon be, widely accessible.

Because of the competitive nature of their industry, telecommunications service providers may be particularly vulnerable to extortion. Persistent loss of service may lead subscribers to go elsewhere, and it is not unknown for a service provider to go out of business after suffering electronic sabotage. Denning (1999: 180–1) describes how one hacker programmed a voicemail system to respond to callers with obscenities, in order to extort free voice-mail boxes from the service provider.

Another case, which illustrates the transnational reach of extortionists, involved a number of German hackers who compromised the system of an Internet service provider in South Florida, disabling eight of the ISP's ten servers. The offenders obtained personal information and credit card details of 10 000 subscribers, and, communicating via electronic mail through one of the compromised accounts, demanded that US$30 000 be delivered to a mail drop in Germany. Cooperation between US and German authorities resulted in the arrest of the extortionists (Bauer 1998). A similar case, reported in January 2000, concerned a hacker somewhere in Eastern Europe who gained access to the customer credit card details of an Internet music retailer, CD Universe, and demanded a payment of US$100 000 (Markoff 2000). After the company refused to pay blackmail, the intruder released some of the credit card details on the Internet. He subsequently boasted that he had identified a flaw in the company's security software used to protect customer information.

The standard security procedures that large institutions usually follow in order to prevent electronic theft may not suffice to ward off intruders. Nicholson (1989) cites a case in England where a person intercepted electronic transaction information transmitted by a bank. He was able to decrypt the data, and threatened to disclose certain information to the Inland Revenue that would embarrass both the bank and several of its customers. The extortionist received £350 000 for his efforts.

Information systems as media for the disclosure of embarrassing personal details

Critics of concentrated media ownership were fond of saying that freedom of the press belongs to anybody who owns one. While ownership of newspapers, not to mention television and radio stations, remains beyond the reach of most citizens, new technologies such as the Internet and World Wide Web provide unprecedented opportunities for the ordinary person to communicate with millions of other people around the globe. The democratization of information technology can be exploited by the extortionist no less effectively than by the ordinary person.

For example, in 1997 two men were charged in the United States with attempting to extort money from fashion model Elle McPherson by threatening to divulge an embarrassing secret and to post two unpublished photos of her on the Internet. In this case the new medium of disclosure was combined with more traditional forms of crime: the threats themselves were communicated in letters, and the photographs allegedly acquired in a burglary (Errico 1997). In a similar incident, a student at the University of Washington broke into the premises of a beverage manufacturer and stole a computer, from which he obtained the recipe for a number of the manufacturer's products. He then contacted the company's president and threatened to post the recipes on the Internet unless the company gave him US$10 000. The extortionist opened a bank account in a false name in order to receive the funds. The threat was never carried out, as the extortionist was identified through photographs taken by the bank's security camera and was also recognized by the daughter of a bank employee (Foster 1998).

In some situations, the threat of disclosure may herald collateral damage. Target organizations may possess information on clients which the clients would prefer remain confidential. At the same time, the organization itself will pride itself on the integrity of its information systems and on the security of its clients' data. Denning (1999: 225) describes how one German bank suffered an attack by a hacker who obtained confidential client information, which he threatened to disclose unless he was paid one million deutschmarks.

Breach of professional–client confidentiality may also border on the extortionate. In 1996, an information technology (IT) consultancy in Sydney allegedly threatened to expose security flaws in a client's system unless it engaged the consultant's services (*Computer Daily News*, 5 June 1996). The following year, a Danish consultant discovered a security flaw in Netscape browser software Navigator 3.0, and threatened to publicize the defect unless the company paid a substantial sum, well above the standard reward it paid for confirmed reports of security holes (Helft 1997). It has also been suggested that some IT consultants have attacked the systems of prospective clients in order to create a demand for their services (Robert Andersen and Richard Hundley pers. comm. 1996). In a sense, this might be regarded as a high-tech equivalent of Munchausen's syndrome by proxy, a psychiatric illness where caregivers injure their charges and then intervene heroically in order to save them.

Information systems as the means of facilitating payment

Traditionally, one of the most difficult challenges facing the extortionist whose goal is financial enrichment has been obtaining payment without drawing attention to his location or identity. The specification of a physical location at which proceeds are to be placed immediately invites surveillance. Even if the threat of harm remains in effect until the perpetrator makes a safe getaway, the presence, however brief, of the perpetrator or an accomplice at a specific location can facilitate identification and/or unobtrusive tracking.

Electronic payment systems now permit funds transfers to locations that may lie beyond the conventional reach of law enforcement. Jurisdictions which are hostile to one's own nation would be less inclined to cooperate in a criminal investigation by assisting in the identification of the account-holder, or in the freezing of assets. Even in the absence of hostile relations between two nations, the very lack of machinery for mutual assistance in criminal matters may impede cooperation to the extent that the offender may elude investigative authorities.

Information systems as incidental to the offence

Just as information technology can be used to facilitate many of the tasks of everyday life, so too can it assist in the more incidental aspects of extortion. The amount of personal information available on the Internet is surprising to many observers. As we shall see in Chapter 10, individual websites can collect extensive data about visitors (Grossman 1997: 188). Denning (1999: 82–3) describes a number of search engines, software tools and other services which enable someone to obtain credit details, driving and criminal records, bank account numbers and financial holdings of a target individual. A skilled extortionist can compile a great deal of information on a target and can use it to frame a very credible threat.

In one ironic case, a police lieutenant in Washington DC who was in charge of investigating extortion plots was arrested and charged with carrying out his own extortion plot against patrons of an establishment catering to a homosexual clientele (Thomas-Lester and Locy 1997). It was alleged that the accused used a law enforcement computer system to identify those who visited the club by matching their automobile licence plate numbers with names on the database. The accused allegedly used another computer in his office to get background information on prospective victims from the Internet.

The use of information systems to develop sophisticated databases for use by extortionists has also been noted in Japan (Sakurada 1994). One Tokyo-based organized crime syndicate, Sumiyoshi-rengo, assembled information on Japanese corporate and political scandals of the previous 40 years, including data obtained from underworld contacts on matters that had never reached public attention. The database is cross-referenced by company name, individual name, and type of scandal. As we have seen, the affiliated criminal organization sokaiya is able, for a fee, to ensure that annual shareholders' meetings go smoothly.

Most extortion attempts entail threats against single targets, whether individuals or organizations. Attempts at what might be termed 'mass extortion', where threats are sent to large numbers of people more or less simultaneously, were fairly rare in years past. Hepworth (1975) cites a late nineteenth-century British case in which three brothers, having received thousands of replies to a published advertisement for an abortifacient, wrote to the respondents over a different signature and from a different address, threatening imminent disclosure and arrest for what was then a crime unless the sum of £2 2s was remitted by postal order. The threat was so convincing that hundreds of remittances were posted to the extortionists, including one from the mother of a healthy baby and one from a woman who did not think she was pregnant but who was afraid that her employer might learn of her concerns (Hepworth 1975: 84–7).

The practice of mass extortion is facilitated by digital technologies. These can help in identifying prospective targets, in communicating the threat, or in inflicting actual harm on information systems. With respect to the identification of prospective targets, it is hardly a secret that a great deal of one's activity in cyberspace can be monitored or traced by third parties. It would, for example, be easy enough to threaten participants in a chat room devoted to 'pre-teen sex' with disclosure to law enforcement authorities, to family members, or to the general public through a web posting. This may entail a degree of entrapment. One could, for example, pose on the Internet as a person (adult or child) seeking an illicit encounter and then threaten to expose willing respondents, as a kind of digital agent provocateur. This contrasts with the threats to inflict bodily harm that characterized the case of the unfortunate Dr Hornbaker.

Computer viruses and cryptography, whose use against single targets was noted above, are also recognizable as instruments of mass extortion. The so-called AIDS InfoTrojan, ostensibly containing information about the AIDS virus, was disseminated by floppy disks to 20 000 people. The disks contained malicious software, which inserted a counter in the user's system start-up files. At the ninetieth start-up, the Trojan encrypted data on the user's hard disk, and brought an announcement to the user's screen directing the user to transfer $US200 to a Panamanian bank account in order to obtain the encryption key (Sieber 1998; Hafner and Markoff 1991: 139).

The Internet and World Wide Web can also serve as sources of instruction for prospective extortionists. One individual who made an extortion threat against Qantas Airways obtained online information on the construction and detonation of explosive devices. Although the threat proved to be empty, the would-be extortionist used the technical details to enhance the credibility of his threat. A later search of his personal computer revealed what appeared to be a script of his telephone threats (Muldoon and Jones 1988).

Controlling Digital Extortion

Adequacy of existing laws

As we have observed, electronic extortion can entail one or more discrete elements, including unauthorized access to a computer, communicating a threat, or damage to a computer system. By now, the substantive criminal laws of most Western industrial societies contain provisions with the capacity to embrace these. Some laws address computer-related extortion explicitly. In the United States, the *Computer Fraud and Abuse Act* (18 US Code Section 1030) specifically addresses the issue of Information Systems as the target of an extortion threat. Section 1030(a)(7) of this Act states that whoever 'with the intent to extort from any person, firm, association, educational institution, financial institution, government entity, or other legal entity, any money or other thing of value, transmits in interstate or foreign commerce any communication containing any threat to cause damage to a protected computer' will be punished (pursuant to 1030(c)(3)(A)) by 'a fine under this title or imprisonment for not more than five years, or both' for a first offense and (pursuant to 1030(c)(3)(B)) by up to ten years' imprisonment for a subsequent offense.

By contrast, the Australian State of New South Wales addresses each of the elements of electronic extortion separately:

> Threatening to destroy or damage property:
>
> s.199. A person who, without lawful excuse, makes a threat to another, with the intention of causing that other to fear that the threat would be carried out:
>
> (a) to destroy or damage property belonging to that other or to a third person; or
>
> (b) to destroy or damage the first-mentioned person's own property in a way which that person knows will or is likely to endanger the life of, or to cause bodily injury to, that other or a third person, is liable to penal servitude for 5 years. (*Crimes Act* 1900 [NSW] Section 199)
>
> Damaging data in computer
>
> s.310. A person who intentionally and without authority or lawful excuse:
>
> (a) destroys, erases or alters data stored in or inserts data into a computer; or
>
> (b) interferes with, or interrupts or obstructs the lawful use of a computer, is liable to penal servitude for 10 years, or to a fine of 1,000 penalty units, or both. (*Crimes Act* 1900 [NSW] Section 310)
>
> Blackmail by threat to publish etc.
>
> s.100A. (1) Whosoever with intent to cause gain for himself or herself or any other person, or with intent to procure for himself or herself or any other person any appointment or office, or with intent to cause loss to any person:
>
> (a) makes any unwarranted demand; and
>
> (b) supports that demand by making:
>
> (i) any unwarranted threat to publish;
>
> (ii) any unwarranted proposal to abstain from publishing; or
>
> (iii) any unwarranted offer to prevent the publication of, any matter or thing concerning any person (whether living or dead),
>
> shall be liable to penal servitude for ten years. (*Crimes Act* 1900 [NSW] Section 100A)

It is not so much the substantive criminal law, but rather the tasks of detection and investigation, which constitute the greatest challenges for the control of digital extortion. As such, and following our theme of legal pluralism, the conventional legal response needs to be considered in the light of other commercial and technological solutions.

Prevention, detection and investigation

Although not widespread, digital extortion is emerging as a threat of some significance. A 1997 computer crime and security survey of 159 Australian businesses found that 10 per cent reported 'extortion/terrorism as the motivation for the unauthorised use of its computer system in the preceding twelve months'; 11 per cent considered that electronic extortion would 'increasingly impact on their organization over the next five years' (Australia 1997). Findings from a similar survey fielded the following year reported 8 per cent of respondents predicting that extortion would become a problem for them (Victoria Police and Deloitte Touche Tohmatsu 1999).

The Canadian Security Intelligence Service (1998) reports that not only individuals but criminal organizations now have the capacity to penetrate corporate information systems. The ubiquity of information systems in advanced industrial societies, and our increasing dependence on them, means that, all else equal, opportunities for electronic extortion will proliferate in years to come. The increasing capacity of prospective electronic extortionists enhances the importance of individuals and organizations protecting themselves against extortion, and for law enforcement and information security industries alike to develop appropriate responses in the event of an attack.

The challenges faced by law enforcement are to identify and locate the extortionist, and take him into custody before he acts on a threat. If this is not possible, the challenge is then to minimize whatever harm may occur as a result of the extortionist's acting on the threat, and to arrest him as soon as possible thereafter. The challenge faced by the victim is to resist the extortionist's demand, and, in most cases, to minimize any adverse publicity occasioned by the extortion. Individuals with something to hide, and organizations with reputations of integrity to protect, may be more concerned to minimize publicity than the victim who is otherwise an attention-seeker. By contrast, the challenge faced by the extortionist is to make a threat which is sufficiently credible, and to achieve the goal of the extortion while escaping detection and identification. Extortionists, like other criminals, will vary significantly in their ability to plan and successfully execute a crime. The less competent offenders will be easier to identify and locate. The perpetrator of the extortion attempt against Bill Gates was relatively unskilled, leaving a number of clues, such as postmarks from locations in northern Illinois and data on the floppy disk, which led to his detection.

One of the more significant concerns is that the technologies of encryption and anonymity, which have begun to foster the move to a cashless economy, will increase the risk that extortion payments will be untraceable. Although the precise architecture of a cashless economy and a global system of electronic commerce have yet to be delineated, suffice it to say that a great deal of thought is being devoted to the challenge of blunting the bad edge of the double-edged sword. Here the solutions reside in code and system architecture, no less than at law. Traceable electronic funds may provide less anonymity than conventional currency does today. Even if a degree of anonymity were to be introduced into electronic funds transfers, there could be provisions for transparency in exigent circumstances. This might include the development of electronic cash systems where the anonymity of the coins could be revoked by a trustee when it becomes apparent that the transaction entails or arises from criminal activity. Such a system could help prevent extortion attacks, which are based on blindfolding protocols, through the involvement of the trustee only at registration of the users (Peterson and Poupard 1997). In some cases, a hashed digital signature will be conclusive proof of having come from a specified terminal and from the holder of the cryptographic key pair. What remains unknown, however, in the absence of more sophisticated user authentication systems, is who precisely is the person using the key pair.

As is the case with other commercial crimes, not all cases of extortion are called to the attention of the authorities. Some victims, particularly financial institutions, are reluctant to reveal their vulnerability, and capitulate quietly. Large organizations in general, particularly those seeking to protect reputational capital, may go to great lengths to avoid adverse publicity. The problems this poses for law enforcement are of some consequence. If a significant proportion of extortions never reach police attention, knowledge of patterns and trends in extortion that might otherwise provide the basis for prevention of attacks and identification of perpetrators will accumulate slowly if at all. The modus operandi of repeat offenders will be that much more difficult to track.

In electronic extortion as in many areas of crime control, one should look first to prevention. Nowhere is this more important than where information systems may be the target of extortion. Effective prevention and control of extortion will require attention to basic principles of management and information security generally. This principle applies across the board, from the individual user to the large organization. Technology and market forces are as important as legal protections.

Individual defences

Because the circumstances of an extortion threat may be incriminating or embarrassing to the victim, there is often a disinclination to report the very existence of such a threat to law enforcement agencies. Short of 'pulling the plug', there is little a person can do to render oneself invulnerable to electronic extortion. One can take basic security precautions, avoid indiscriminate disclosure of personal information, be cautious with one's PIN and credit card details, and report apparent irregularities to one's ISP and to appropriate authorities.

Organizational defences

In information security, redundancy has its virtues, and to the extent that it is feasible, a backup of systems will be useful to neutralize a threat. Basic principles of information security are now enshrined in standards such as BS7799 in Britain and AS4444 in Australia; adherence to these is necessary, but not sufficient, to minimize extortion risk.

Organizations that might be vulnerable to extortion would be well advised to undertake a risk analysis across all aspects of their operations. At the most general level, this will entail analysis of system vulnerabilities, especially those relating to accessibility by the outside world. Basically, the vulnerability of an information system to extortion, as it is to other forms of attack, will vary inversely with the quality of systems administration, the degree of rigour in access control, and the superiority of its organizational culture. Cultural factors may be as important as individual threats, since an apathetic or uncaring workforce may be less likely to warn management of suspicious activities.

Threats are by no means limited to external sources. The possibility that a current or recently disengaged employee of the organization, alone or in collaboration with an outsider, might seek to attack the organization's resources, means that great care should be given in the engagement, and disengagement, of employees. Restructuring, downsizing, outsourcing, engagement of temporary staff may erode individual loyalties or inflame individual grievances. This may be of particular significance when aggrieved individuals have, or have had, access to information systems (see the case of *R* v *Stevens* unreported decision of the NSW District Court, 27 March 1998, referred to in Chapter 2). Relations with employees are an important part of an organization's risk profile, and vulnerability to a range of employee criminality can depend on an enlightened human relations program.

This is easier said than done, as the changing nature of organizational life may offer unprecedented opportunities to potential extortionists. The exceptional mobility that characterizes employment in many industries today means that bonds of loyalty to an organization are often weak. Nowhere is occupational mobility more apparent than in the present-day information technology industry, where terms of employment may be counted in months rather than decades. To the extent that organizations are dependent on external technical expertise, through IT outsourcing or through contact at the time of installation with manufacturers' representatives, the importance of ensuring the integrity of suppliers is that much greater. Some chilling anecdotes have begun to accumulate about criminals posing as IT consultants and obtaining crucial information about a target system by interviewing an organization's senior IT staff.

Whatever the source of threat, good IT auditing practices are an important defence against extortion, as they are against other misdeeds. Many full-blown attacks are preceded by apparently trivial intrusions for the purpose of 'probing' a target system (Sommer 1999). Organizations may insure against losses from extortion, both 'high tech' and 'low tech'. The availability and terms of extortion insurance will depend on the degree of security that an organization has in place. It may well be a condition of such insurance that its very existence be kept confidential, lest prospective extortionists be led to the conclusion that the target organization may be able to shift the costs of an extortion to an insurer, and be that much more inclined to capitulate to an extortion demand.

Organizations should also have contingency plans in the event that the above measures fail to prevent an extortion attack. Just as many organizations have emergency plans in case of fire or electrical failure, so too they can plan for criminal attack. The prospect of responding to an incident without any prior thought being given to appropriate procedures should be chilling to anyone. Areas for planning include the designation of a crisis management team, relations with law enforcement agencies, and the management of media disclosure. Response to an attack may require external IT expertise, the source of which should ideally be determined in advance. The prospects of facing a crisis without any established procedures should be daunting to any executive. One cannot imagine having to leaf desperately through the Yellow Pages in search of a cryptologist.

Among the first tasks to be faced is to determine the plausibility of the threat. One needs to know whether the threat is tangible (an actual cryptovirus in place, as opposed to a veiled threat of prospective system damage). It is also important to know whether the threat can be located and neutralized or whether one's operations can continue on a backup system.

Investigation

The understandable reluctance of some victims of extortion to disclose their predicament stands in contrast to the public interest in mapping the distribution of extortion over time and space. Extortionists may be serial offenders, and as is the case with conventional criminal activity, evidence left at one crime scene may enhance the investigative capacity of the police. It is therefore no less desirable in circumstances of extortion than in other forms of electronic theft that victims be encouraged to report incidents. Discreet management of this information by public authorities is therefore essential.

The initial approach by an extortionist may provide an opportunity for early identification. As is the case with ordinary extortion attempts, investigators of cyber extortion will seek to prolong the dialogue between the target and the extortionist in order to establish the identity and location, and to mobilize other countermeasures.

The anonymity of electronic payments systems would facilitate concealment of the recipient of the proceeds of extortion. When extortion payments are met with ordinary electronic cash transfers, the destination will be identifiable. The extent to which banking and law enforcement officials in the host jurisdiction will be of assistance will depend on that nation's bank secrecy laws and the relationship between the law enforcement agencies of the two jurisdictions.

Conclusion

As we noted at the beginning of this chapter, extortion is hardly a new phenomenon. Indeed, its rudiments are very likely prehistoric. But the digital age offers many new opportunities for extortion and poses many new challenges for its control. The proliferation of information technology has brought about new media for the communication of extortion threats, and new targets. But security systems can be put in place to shield electronic assets, at least. Moreover, technologies of tracking and interception have placed new tools in the hands of those who would investigate digital extortion, and the growing capacity for international cooperation will help contain its activities between jurisdictions, as well as within them.

CHAPTER 4

Defrauding Governments Electronically

Throughout history attempts have been made, often successfully, to defraud government agencies. In the early eighteenth century, Robert Loggin, an English Customs official, conducted an extensive analysis of fraud relating to trade in pepper and tobacco and estimated that over £1 million had been lost to the government in the years 1715 to 1717 alone. He concluded:

> He therefore, who defrauds the Customs, does not only wrong the government; but commits a particular injury against every member of the community. To make this evident, I need but observe, that the man who in trade vends his commodities, without paying the duties, must either cheat the buyer by his exorbitant gains, or ruin the fair trader, who complies with the laws (Loggin 1720: 57–8).

Other examples of fraud abound throughout history – from ancient Greece and Rome to the early courts of England and France in the seventeenth and eighteenth centuries (see Johnstone 1999 for a review). In each case, governments were targeted as they provided a ready repository of funds, and since crime follows opportunity, fraudsters sought to take advantage of what was on offer. Some of the largest losses that governments have traditionally sustained relate to the evasion of payments due to them, such as taxes and fees, and obtaining benefits to which the recipient is not entitled, such as welfare benefits and educational and travel allowances.

Fraud may also be perpetrated by insiders. Government employees have stolen funds and other government property both directly and indirectly. Direct theft may occur when employees steal petty cash or remove government property. More covert forms of theft involve the abuse of government facilities such as the unauthorized use of motor vehicles and computers. Government employees are also well placed to abuse their position by accepting bribes to grant licences for which there is no entitlement or to charge governments for goods or services that are not in fact provided (see Mills 1999).

Difficult questions arise as to the proper characterization of such acts and whether they can be described as crimes of theft, or merely 'leakage' of government resources due to poor internal controls. The scale of such conduct also varies considerably from the trivial, such as having an extended lunch break, to the serious, such as large-scale revenue fraud.

Little systematic research has been undertaken into the nature and extent of the losses governments have sustained. Although some, but by no means all, agencies record information on the extent of fraud for their own internal fraud control purposes, they rarely share it publicly. Often all that is known is what is mentioned in brief summaries in annual reports or media reports of cases involving prominent figures. Many governments would prefer that their fraud experiences never be made public in order to avoid criticism for not having appropriate preventive measures in place. There are, however, some limited indications of the extent to which governments have been defrauded.

In KPMG's survey of the fraud experiences of a sample of large Australian organizations, 62 per cent of the 39 government agencies surveyed had experienced some form of fraud in the preceding two years (KPMG 1999). In another Australian survey conducted by Deakin University for the Victoria Police Major Fraud Group, the types of fraud most often experienced by the 40 public administration agencies surveyed involved misappropriation of cash, stock and equipment, unauthorized use of government equipment, and false or inflated claims on travel or expense accounts (Deakin University 1994).

In recent years, new forms of fraud are beginning to emerge which make use of information and communications technologies. Because of the extensive use they make of computers, governments have been frequent targets; as government services continue to be provided through online facilities, so will the opportunities for fraud increase, potentially with profound consequences.

In the United Kingdom, for example, the government expects 25 per cent of government services to be available electronically by 2002 (UK Prime Minister 1999), while in Australia the Federal Government has determined to provide all appropriate government services online by 2001 (Attorney-General's Department 1998). As many of these strategies entail the widespread use of electronic payment systems, the potential for fraud is substantial, as we have seen in Chapter 2.

In Australia, state and federal governments are well advanced in the implementation of the electronic delivery of services and could be said to be leading the world in this area. The Federal Government, for example, has developed a strategy, Project Gatekeeper, which aims to provide a common platform for the development of systems that rely on public key cryptography and digital signatures. This strategy, which is currently being revised to accommodate various suggestions for improving the operation of the platform, seeks to provide a system of secure electronic communications on public networks when dealing with the government (Australia, Office of

Government Information Technology 1998). In the State of Victoria, an online government service known as 'Maxi' has been created which enables members of the community to gain access to Victorian Government and certain private sector services through the telephone, the Internet and at community kiosks (Victorian Government 2000). 'Service New South Wales' provides a similar service, while in the ACT people are able to obtain government and community information and pay certain bills at 'Austouch' kiosks (ACT Government 2000). In South Australia, it is now possible to pay for motor vehicle registration, water accounts, and certain other services online. The use of such technologies will undoubtedly increase the extent to which governments are able to discharge their responsibilities to the communities they serve both efficiently and cheaply.

The possibility exists, however, that dishonest individuals might seek to misuse computers to defraud governments or otherwise to steal from them. Already this is taking place. In a survey of computer crime and security conducted by the Australian Federal Government Office of Strategic Crime Assessments and the Victoria Police Computer Crime Investigation Squad (1997: 30), 36 per cent of the eleven government agencies surveyed reported misuse of their computer systems; 45 per cent reported external forms of attack, that is intrusion accomplished through remote access to computer systems. The most common types of computer abuse reported by the government agencies surveyed related to damage or unauthorized access to, or copying of data and programs.

In Britain, a survey conducted by the Audit Commission of 5500 public and private sector agencies in 1994, found that of the 24 government agencies which responded, 46 per cent had experienced some form of computer abuse in the preceding three years. Of the 66 incidents reported, 21 involved fraud amounting to £678 874 in total (Audit Commission 1994).

In this chapter we consider the opportunities that exist for crimes of acquisition to be perpetrated against government agencies through the use of computers and communications technologies. We shall consider conduct that can be defined as crimes of acquisition, in the most general sense of the term, including conduct that would result in a successful prosecution for theft, through to less serious forms of improper use of government resources and property. It is appropriate to consider the full range of such instances, as often the most seemingly trivial incident can indicate more widespread problems (see Stoll 1991). The discussion will conclude with a brief review of some of the more important strategies that may be used to control electronic fraud against governments. As in previous chapters, the identification of such strategies will reveal the importance of a pluralistic approach that makes use of practical fraud prevention as well as traditional legal measures. Controlling electronic fraud against governments will also require the mobilization of private sector resources in addition to government-based initiatives. In this area in particular, Lessig's (1999) assertion that regulation by 'code' needs to accompany regulation by law is particularly apt.

The Nature and Extent of the Problem

How then are governments being victimized through the use of computerized technologies, and what opportunities exist for such technologies to be used for financial crime in the future? The vulnerability of governments falls into five categories which are listed in order from those that involve the largest financial losses to those that involve the least: theft of benefits, money, information, computer hardware and software, and time. Although this arrangement in terms of importance is not based on empirical evidence, it does give some subjective indication of the areas of greatest concern.

Theft of benefits

Revenue fraud

Revenue departments throughout the developed world are now making extensive use of information technologies in the assessment and processing of private and business taxation liabilities. Already some revenue departments permit individuals to lodge income tax and value-added tax returns electronically, while refunds may be paid through the use of electronic funds transfers. As global online commerce increases, the collection of revenue will be greatly facilitated by the use of computers, and it may become obligatory for businesses to deal with revenue departments electronically in order to expedite the processing of claims and to facilitate computerized data-matching for risk management purposes. In Australia, for example, a system for the collection of goods and services tax (GST), which began operation on 1 July 2000, requires all businesses to communicate electronically with the tax office if their annual turnover exceeds A$20 million. If electronic GST returns are given they must be in a form approved by the tax office and secured by an electronic signature (ss. 31–25 *A New Tax System [Goods and Services Tax] Act* 1999 [Cwlth]).

The collection of revenue electronically, however, creates various security risks. Most relate to attempts to disguise transactions in order to avoid paying tax, particularly consumption taxes such as value-added taxes and goods and services taxes (see Bridges and Green 1998). In addition, attempts may be made to manipulate the payment of refunds or to increase entitlements to benefits. Manipulation of the input tax claiming system is a well-recognized way of defrauding the government in a value-added tax system, and where information is mainly processed electronically, digitally fabricated documents could be submitted in support of illegal claims for input tax.

Taxpayers might also seek to disguise their true identities in order to avoid the collection of unpaid taxes and penalties. One of the greatest risks lies in the process of registering taxpayers electronically. Although the use of electronic registration may expedite the processing of registration, this could present security risks if original documents are not subjected to appropriate scrutiny. It is also easier to disguise one's true identity through the use of electronic communication technologies than by conducting business face to face. If adequate proof of identity is not obtained and if evidence of identity is not properly verified with the issuing source, it may be possible to perpetrate a variety of fraudulent activities. Such illegality may be facilitated where tax office employees provide access to networks or conspire with outsiders to overcome security systems. Already there have been reports of Australian Taxation Office employees having been involved in fraud and corruption. In the State of Victoria, for example, 242 instances of illegality were reported between January 1994 to December 1997 (Hughes 1998).

In 1994, the UK National Audit Office conducted a review of HM Customs and Excise in order to determine the extent of fraud carried out by personnel within the organization. Between 1989 and 1993, 122 confirmed or suspected cases of internal fraud were dealt with involving £205 000. Ninety-seven cases of internal fraud amounting to £164 000 were proved. In some cases VAT payments and refunds were diverted to individuals within Customs, while other cases involved corruption in which officers cancelled VAT liabilities in return for cash payments (UK National Audit Office 1994). The use of electronic claiming and payment systems would clearly facilitate fraud of this nature.

Prosecutions have also been taken against individuals who defraud taxation authorities by selling computers without paying sales tax. In Victoria, one person was sentenced to a suspended term of imprisonment for failing to pay $1.143 million in sales tax on computer equipment (*DPP [Cwlth]* v *Thomas* [Court of Appeal, Supreme Court of Victoria, 28 August 1997]), while another pleaded guilty to having failed to pay $900 000 in sales tax. In Brisbane, an offender was sentenced to two years' imprisonment with a minimum of eighteen months for having avoided sales tax on computer sales amounting to $113 000 (*AFP News*, December 1998). The Australian Taxation Office expects to recover $100 million in unpaid sales tax on computer equipment following approximately 500 prosecutions in recent times (Sinclair 1999b).

One final area of concern regarding loss of government revenue lies in the problem of taxing digital products and services themselves. Although this is arguably outside the concept of electronic fraud as considered in this book, loss of revenue through the importation and sale of digital products and services is likely to be of considerable importance in the years to come. Several possibilities have been canvassed concerning the recovery of taxation on products imported electronically, such as through the Internet. The main problem is the identification of a suitable organization to collect taxation as the only controlling entity would be an Internet service provider. Other suggestions have been raised in Europe for financial institutions to be responsible for collecting taxation when transactions are cleared electronically (Jenkins 1999). These matters are, however, outside the scope of our present inquiry.

Customs and excise fraud

Customs and excise duty may be lost through the manipulation of computerized cargo management and revenue collection systems. In Australia, for example, more than 98 per cent of the 3.7 million import and export entries were cleared electronically by the Australian Customs Service in 1997–98 (Australian Customs Service 1998). This creates considerable opportunities for electronic manipulation of the system by reducing the number or value of goods imported in order to reduce liability for duty, or by creating bogus exports in order to support drawback claims.

Although agreement has not been reached in many countries on the imposition of tax and duty on the importation of digital products, such as software downloaded from the Internet, the potential loss of revenue here could also be substantial in the future. In Britain, the Chartered Institute of Taxation has warned of the potential loss of billions of pounds of revenue as a result of cyberspace trading (Eden 1997: 178, n. 80). A related area of concern in terms of loss of revenue on transactions based on electronic commerce is that there will be an increase in trade between overseas suppliers and private consumers, thus excluding wholesalers and retailers. Collection of value-added taxes and duty from end-consumers probably represents the most difficult challenge in terms of achieving compliance since there is no internationally agreed standard for auditing digital transactions. Even proposals to link tax and duty collection to electronic commerce payment systems (e.g. Microsoft and Visa's 'SET') have encountered problems through the absence of international agreement on the protocols to be adopted (Jenkins 1999). As Braithwaite and Drahos (2000) argue, it is to be hoped that the future will see enhanced global mechanisms develop for business regulation in which world citizenship may play a leading part.

Finally, in the future, customs controls on the movement of currency may also be infringed where digital cash is brought into or taken from the country electronically, such as through the use of smartcards. This, of course, poses particular concerns in terms of money laundering where financial institutions are not involved at all in electronic transactions carried out using smartcards and value is able to be moved internationally without detection by cash transaction-reporting agencies (see Grabosky and Smith 1998).

Social security fraud

Fraud directed at social welfare benefits has traditionally been perpetrated by those who might be described as the needy and the greedy – that is, people who could be said to have acted out of genuine financial need as well as those acting purely for personal gain. The aged and women are often involved; women, in particular, are proportionally overrepresented in fraud statistics largely because of their involvement in social security fraud (Daly 1989; O'Brien 1991; Wilke 1993).

As government benefits programs continue to be administered electronically, the opportunities for electronic social security fraud by both the greedy and the needy are also enhanced. Already electronic benefits transfer (EBT) is being used in the United States and in Australia, and unfortunately has been subject to abuse. EBT was initially trialled by the US Government in 1984 when its Department of Agriculture tested the viability of the system as a substitute for the delivery of paper food stamps (Warton 1999). It currently operates in most American states, and following the *Welfare Reform Act* 1996, all states will use EBT by 1 October 2002 (Warton 1999). In Australia, EBT has been used for the delivery of social security benefits by Centrelink since 1997 in several major Australian cities. The system operates on a national scale and assists in the electronic delivery of limited social security benefits in cases previously addressed using the traditional counter cheque. Plastic cards can be used to obtain cash from ATMs by authorized recipients.

As long as social security recipients have access to the necessary technology, they will be able to steal from governments electronically. The possibility also arises that those who maintain and operate online payment systems within agencies may be tempted to act illegally by diverting funds intended for welfare recipients to their own purposes.

In Australia, a number of prosecutions have resulted from internal fraud carried out by government employees fraudulently using the EBT computer system in December 1997 and January 1998. EBT cards were issued in fictitious names, enabling the offenders to obtain cash at ATMs. In one case, the proceeds of the fraud were used to purchase heroin in the same street as the location of the ATM and within ten minutes of the fraud occurring (Warton 1999; *AFP News* December 1998).

Solutions to the problem of social security fraud may also make use of electronic technologies. In Connecticut, in the United States, for example, fingerprint scanners were introduced in 1996 in order to prevent social security fraud. Recipients of welfare cheques were required to undergo fingerprint scanning before collecting their payments. The introduction of the technology cost US$5.1 million but is said to have saved the state US$9 million in fraudulent claims (Denning 1999: 324).

Health benefits fraud

Because of the substantial sums of money which governments allocate to public health care, those who provide services and those who receive payments may be tempted to act fraudulently. In the United States, Medicaid/ Medicare fraud has been considerable, with the risk of detection minimal and the potential gains substantial (Jesilow, Pontell and Geis 1992). Manipulation of electronic claiming and payment systems has been in existence in that country for many years now; precise estimates of losses sustained are not available, but there is some anecdotal evidence. In November 1988, a medical practitioner and his two sons were convicted in New York of stealing US$16 million from Medicaid between 1980 and 1987. The offenders falsely billed Medicaid for approximately 400 000 phantom patient visits by programming their medical centre's computer to generate some 12 000 fictitious patient consultations a month (Sparrow 1996: 130–1). Often computerized data analysis tools are unable to detect such fictitious claims as they have all the characteristics of real claims.

In Australia, the Health Insurance Commission (HIC) processes claims and makes payments for the provision of health services and other benefits under various government programs. Many transactions are now dealt with electronically, which creates substantial risks of misappropriation of funds and fraud by reason of the large sums of money involved. Between 1 July 1997 and 30 June 1998, for example, 128 023 Medicare services amounting to $7 461 353 were processed by electronic funds transfer which, though a relatively small proportion of the 202.2 million Medicare services billed worth $6334 million in the same year, will increase considerably in the future.

At present, the most common offences investigated by the Commission relate to claims for Medicare or Pharmaceutical benefits being made by means of false or misleading statements. For the year 1996/97, the HIC's *Annual Report* indicates that $3.9 million of benefits paid incorrectly were recovered or were being recovered from providers and the public. For the same year, a total of 2130 complaints of alleged fraud and inappropriate practice were recorded on the HIC's National Complaints Register (HIC *Annual Report 1997–98*, Professional Review Supplement: 17–18).

As the use of electronic processing of claims increases, so the risk of fraud also increases, perhaps facilitated by the computer technologies adopted. Risks relate to the possibility of electronic claim forms being electronically counterfeited, or manipulated, signatures being forged, and electronic funds transfers being altered or diverted away from legitimate recipients (Smith 1999a). The HIC has already been subject to fraud perpetrated by insiders and the possibility exists that those with the technological skills could attack the Commission's electronic claiming system internally. In 1997, for example, two former HIC employees were convicted of defrauding the Commonwealth by creating false provider accounts and making illegal claims to the combined value of more than $45 000 (HIC *Annual Report 1996–97*, Professional Review Supplement: 23).

Credit card fraud

Government funds may also be misappropriated through the dishonest and unauthorized use of government credit cards. Providing government employees with credit cards is an efficient and secure way of paying for authorized goods and services, as funds need not be drawn in cash, which could be stolen before use. A survey undertaken by the Australian Department of Finance and Administration found that in the year to April 1998, there were 11 287 Australian Government Credit Cards in use that were used for 484 000 transactions with a total value of $162 million (Joyce 1999).

Transacting government business through the use of plastic credit cards, however, raises all the security risks that cards bring with them (see Smith 1998b). More important, however, is the possibility that government employees may use cards for unauthorized purposes. In 1994, the Australian National Audit Office conducted an audit of a sample of transactions undertaken with the Australian Government Credit Card (Australian National Audit Office 1994). Since the card was introduced in November 1987 until March 1994, there were 46 cases of fraud reported totalling between A$1.8 million and A$2.0 million for all departments and agencies. The bulk of cases related to claims under A$5000, with most of the frauds relating to the unauthorized purchase of goods to be used for private purposes or for travel and hospitality which had been paid for from other sources ('double dipping'). Of a sample of 1866 card transactions examined, the Australian National Audit Office identified 523 instances of misuse of cards, some not formally reported. These instances related to 336 transactions in which the use of the card was not correctly approved, where use occurred outside prescribed guidelines, or where use was questionable or inappropriate.

Despite widespread publicity about the risk that improper use of official credit cards will be detected, government employees continue to abuse them. The result is often loss of employment and criminal prosecution when the abuse is uncovered (see Forbes 1999). The Australian Civil Aviation Safety Authority, for example, recently identified two cases in which its officers had abused Travelcards issued for official business. One officer had withdrawn his travel allowance from the Authority's bank and then used his Travelcard to pay for accommodation, meals, drinks, and in-house videos while on official business. Another used the Travelcard as a form of personal credit line by withdrawing cash and repaying it at a later time that was convenient (Joyce 1999).

Theft of telecommunications services

Government operations throughout the globe rely to a great extent on the use of telephones and, as such, fraud that is directed at telecommunications systems has the potential to inflict considerable losses. There are innumerable ways in which to manipulate telecommunications systems in order to get services without having to pay for them (see Chapter 5 below, and Grabosky and Smith 1998: Ch. 4), the most recent of which entail the use of digital technologies to compromise the security features of cellular telephones. When carriers were wholly owned by governments, losses remained in the public sector. Now, such losses are largely sustained by private companies.

Government agencies and their staff, however, continue to make considerable use of telecommunications and will occasionally incur substantial losses through theft and fraud, particularly that involving PABX technologies. In 1999, the UK Treasury identified a number of instances in which telecommunications services had been stolen from public sector agencies. One case of PABX fraud cost a department the equivalent of US$80 000 a day, while another department lost US$640 000 over six weeks (Wraith 1999). In yet another case, illegal access was gained to Scotland Yard's PABX system in London by computer hackers based in the United States; unauthorized international calls to the value of A$1.29 million were made for which Scotland Yard was liable (Tendler and Nuttall 1996). Other cases have involved teeing into government telephone lines and cloning of cellular telephones, many of which have involved employees within organizations disclosing confidential security information to enable the crime to be committed (Denning 1999: 177–83).

Theft of money

Misappropriation of salaries and payments

Manipulation of computerized payment systems has been used for decades as a way of stealing from government agencies. An early case in the United States, for example, involved an employee of a welfare department who stole US$2.75 million over nine months by entering fraudulent data into the department's computerized payroll system. He then intercepted salary cheques sent to the phantom employees, endorsed them and cashed them. The fraud was uncovered when a police officer noticed some of the cheques in the offender's illegally parked rental car (Brandt 1975).

The Australian Federal Police have also investigated a number of instances where individuals have made use of federal computers in Australia to divert funds from government accounts. In one case, a programming contractor altered a government department computer program so that funds would be automatically transferred to the individual's personal bank account (Baer 1996: 24). In the Australian Capital Territory (ACT) in 1998 a financial consultant to the Department of Finance and Administration allegedly transferred $8.725 million electronically to private companies in which he held an interest, after logging on to the Department's computer network using another person's name and password. The individual in question was charged with defrauding the Commonwealth. It was reported that $5.48 million had not been recovered by the police after an investigation (Campbell 1999).

Counterfeiting and forgery

Governments may also be victimized through acts of forgery and counterfeiting carried out through the use of desktop publishing equipment. Counterfeiting currency issued by central banks has been greatly facilitated by electronic scanning equipment and colour photocopiers (Baer 1996). These technologies have also been used to forge government cheques, benefit claim forms, and payment vouchers along with primary and secondary documents used to establish false identities in connection with criminal activities.

Counterfeiting is, of course, not new. Daniel Perrismore, for example, was fined and pilloried for forging £100 notes in England in 1695 (Rastan 1996) and since then elaborate measures have been taken to improve the security of currency. Even Australia's polymer substrate currency, reputed to be one of the world's most secure and protected through the use of clear windows and holograms (James 1995), continues to be forged, with convincing copies being produced electronically. In Western Australia, for example, in November 1998, four school students were allegedly involved in the counterfeiting of A$50 notes by means of computers, scanners and colour printers. Another case in Perth allegedly involved a 23-year-old university student who had used computer systems to counterfeit A$100 notes (*AFP News* December 1998).

In Britain, currency counterfeiting increased more than 250 per cent between 1991 to 1994, with almost ten per cent being produced by colour photocopiers. While such notes are easy to detect, more sophisticated counterfeiters use paper stocks and dyes which more closely resemble the original and efforts are made to fabricate the other security features of the original document (Rastan 1996).

The introduction of the Euro as the single currency in eleven European Monetary Union countries, which is due to commence on 30 June 2002, may also increase the risk of counterfeiting as the new currency will be unfamiliar to many citizens of member states. Concerns have also been expressed that differences may exist in the quality of the new notes, which are to be printed in different locations (Hook 1998). In addition, the Euro will be printed in 100, 200, and 500 denominations, the largest of which might be more attractive targets for counterfeiters and money launderers.

In Canada in 1996, about 70 800 counterfeit banknotes worth approximately Can$1.4 million were detected throughout the country. With over 800 million notes in circulation worth approximately Can$29 billion, the proportion of counterfeits is relatively low. As in other countries, however, the use of computerized technologies to counterfeit notes has resulted in a large increase in the rate of illegal note production. The government's Central Bureau for Counterfeits now makes use of a sophisticated database for the notification and tracking of counterfeit notes (Duncan 1998).

Between 1987 and 1996, the volume of counterfeit United States currency increased from US$89 million to US$205 million, peaking at US$339 million in 1995. Although the new US$100 note is more difficult to counterfeit than earlier versions, the use of desktop publishing equipment to counterfeit United States currency has increased substantially. In 1987 it accounted for 34 per cent of counterfeits, while in 1996 82 per cent of counterfeits were computer-generated (Johnson 1997).

Theft of information

Government employees have also stolen information by obtaining unauthorized access to computers, copying data, and selling the copies to third parties (Baer 1996: 24). For example, charges were laid against an employee of the Australian Department of Social Security after the removal of a large quantity of records from the Department's database. Details of individuals held on the database were sold to a private investigator who sold them on to insurance companies (Australian Federal Police 1996: 20). In 1996, an employee of the DSS, a former police detective, was sentenced to 200 hours' community service and fined $750 in Sydney after he was found guilty of unlawfully gaining access to and disclosing DSS information (Australian Federal Police 1997: 30).

Intellectual property

Government employees have access to and make use of various forms of intellectual property in connection with their employment. Copyright, patents, trademarks, designs and certain other specific rights may all be used without authority. The greatest area of risk, however, lies in government-owned software being downloaded and used on personal computers for private purposes. This form of illegality is not, however, confined to the public sector. The American Society for Industrial Security's latest Trends in Intellectual Property Loss Survey, for example, found losses of US$44 billion over seventeen months in the United States, considerably more than the US$5.1 billion estimated to have been lost in its 1996 survey. The Society estimates potential losses for all United States companies to be more than US$250 billion annually.

Espionage

The possibility also exists that government employees may sell confidential, sensitive information obtained in the course of their employment. Although this has always been a risk, particularly in matters involving national defence, the use of computers to discover information, such as through hacking, or to transmit information obtained illegally, makes the problem potentially much worse. In the United States there have been numerous incidents of government employees selling intelligence information to foreign governments. In one case in 1997, a high-ranking agent working for the CIA had attempted to locate sensitive information within the agency's computerized databases by carrying out searches for the keywords 'Russia' and 'Chechnya' and trying to gain access to databases without permission. After an investigation, the FBI seized his computer notebook and found classified CIA documents on its hard drive and a floppy disk containing summary reports of CIA human assets. He admitted to selling top-secret intelligence information to the Russians for US$180 000 (Denning 1999: 133).

Computers have also been used for economic espionage. In Australia, some years ago, the NSW Independent Commission Against Corruption (ICAC) (1992) investigated employees of the national telecommunications carrier, Telecom, who had sold confidential government information to private investigators.

Theft of computer hardware and software

In the private sector, theft of computer hardware and software represents a considerable problem for businesses. In one survey undertaken between

September 1997 and June 1998 by the Rand Corporation of 95 high-technology businesses in the United States, theft of high-technology components was estimated to cost more than US$5 billion a year in direct and indirect costs, amounting to approximately 2 per cent of industry sales. The study also found that there had been a gradual reduction in these losses because of businesses making use of various theft prevention strategies (Gengler 1999; Rand Corporation 1999).

Government employees are also in a position to steal computer equipment, which often contains valuable software and sometimes sensitive information. In one recent investigation undertaken by the Australian Federal Police, a government department was the victim of a series of thefts of computers containing sensitive information. A number were recovered and three people charged. The department in question has since undertaken a review of its security and employee screening procedures to prevent similar incidents (Australian Federal Police 1998: 43).

Laptop computers are particularly attractive targets for thieves, not only because of their portability, but also because of the information they hold. One computer insurance company in Columbus, Ohio, received 309 000 claims in respect of stolen laptop computers and 100 000 claims in respect of desktop computers in 1997, worth US$1.3 billion in all. Where data held on hard drives are not backed up, considerable inconvenience can arise from a theft. United Nations officials reported, for example, that the theft of four computers from its New York offices, which contained data on human rights violations in Croatia, resulted in delays in the prosecution of war crimes from this region (Denning 1999: 157–8).

Theft of time

Finally, one of the most difficult types of computerized theft from the government to detect is the theft of time (and, incidentally electricity) when employees use computers for unauthorized personal use at times when they should be carrying out legitimate work. A related consequence of inappropriate use of computers is that if communications bandwith is used for non-work-related activities, this may slow down or prevent the network from being used for legitimate activities. One can imagine, for example, the problems that would be created for a government departmental server if staff regularly downloaded movies electronically.

Misuse of government resources in this way results in considerable loss of productivity as well as creating an inappropriate culture in the workplace and potentially leading to problems of legal liability. The greatest area of vulnerability in recent times relates to inappropriate use of the Internet. Compaq Computer Corporation, for example, found in an analysis of the Web sites visited by its employees that twenty people had visited more than 1000 sexually explicit sites each in less than a month. Another company found that 15 per cent of Web sites visited by its employees contained sexually explicit material. In a survey of executives conducted by *PC World*, 20 per cent of the 200 companies surveyed had disciplined staff for misusing Internet access; the most common offences involved visits to pornographic Web sites, shopping, using chat lines, gambling, and downloading illegal software (Denning 1999: 360–1).

Misuse of the Internet in this way also occurs within the public sector. In Australia, the operator of Internet sites that provide live, online sex shows conducted a survey of her users and found that during business hours between July 1999 and June 2000, 1059 people who connected to the sites had used 'gov.au' domain names. They came from a range of federal, state, and local government offices located throughout Australia. It is claimed that some 4000 individuals have used government-provided facilities to gain access to sexually explicit sites (see Sinclair 1999a; Taylor 2000).

Preventive and Control Strategies

A wide range of strategies have been developed to prevent traditional forms of public sector fraud which range from the creation of guidelines and policies on fraud control to the use of computer-based security techniques. Each is able to be applied to the realm of electronic theft, some with greater effect than others. The following strategies are presented in groups which range from those that arguably entail the least expenditure and that are likely to yield the greatest benefits, to those that are more expensive but less likely to be effective. Once again, this arrangement is largely subjective and made in the absence of the results of any quantitative research into costs and benefits.

Management of fraud control

Most government agencies throughout the world have detailed fraud control policies in place which provide guidelines on the establishment, implementation, and management of agencies in order to reduce fraud risks. There has also been a recognition in recent times of the need to create an ethical environment in the public sector by educating public servants about the desirability of complying with laws and codes of practice. In this sense, public servants may be seen as standing in a fiduciary relationship to the community such that their conduct should be governed by an overriding duty to act in the best interests of the community (see Mills 1999).

In 1987, the Australian Federal Government embarked on a review of its systems for dealing with fraud, and concluded that each agency should have primary responsibility for dealing with fraud, that the emphasis should be on fraud prevention rather than deterrence-based approaches, and that agencies should bear primary responsibility for investigating minor instances of fraud. Procedures were then introduced to implement these reforms; these now comprise the Fraud Control Policy of the Commonwealth, which was substantially revised in 2000.

Of particular importance is the need to develop specific policies on computer security along with appropriate guidelines on reporting computer misuse and abuse. The Commonwealth of Australia's Fraud Control Policy now specifically includes such matters. Many jurisdictions also now have public interest disclosure legislation which aims to ensure that those who report illegal conduct are not disadvantaged by their behaviour. In the case of computer-based illegality, as in other areas of crime, severe penalties could be imposed on people who engage in, or attempt or conspire with others to carry out acts of reprisal against those who disclose illegality in the public interest. To date such remedies have rarely been used.

Policies also need to deal with specific online behaviour of employees. Agencies should establish guidelines, for example on access to and use of the Internet for private purposes, personal use of electronic mail, downloading government software, and the use of copyright material. In Australia, the Federal Privacy Commissioner has recently issued *Guidelines on Workplace E:mail, Web Browsing, and Privacy* which recommend steps that organizations can take to ensure that their staff understand the organization's position on these matters through the development of clear policies. The guidelines seek to ensure that management spell out clearly their expectations and permitted practices to employees (Office of the Privacy Commissioner 2000).

Personnel monitoring

One of the most important areas in which technology-based fraud against governments can be curtailed lies in ensuring that trustworthy and reliable staff are employed, particularly in senior positions of responsibility. The administration of modern technologically-based security systems involves a wide range of personnel from those engaged in the manufacture of security devices to those who maintain sensitive information on passwords and account records. Each has the ability to make use of confidential information or facilities to commit fraud or, what is more likely to occur, to collude with people outside the organization to perpetrate an offence. Preventing such activities requires an application of effective risk management procedures within agencies which extend from pre-employment screening of staff to regular monitoring of the workplace. Long-term employees who have acquired considerable knowledge of an organization's security procedures should be particularly monitored, as it is they who have the greatest knowledge of the opportunities for fraud which exist and the influence to carry them out.

Computer usage monitoring

Employees' use of computers and their online activities can be monitored through the use of software which logs usage and allows managers to know, for example, whether staff have been using the Internet for non-work-related activities. Ideally, agreed procedures and rules should be established which enable staff to know precisely the extent to which computers are able to be used for private activities, if at all. If agencies do permit staff to make such private use of computers, then procedures should be in place to protect privacy and confidentiality of communications, subject, of course, to employees obeying the law.

Where certain online activities have been prohibited, many government agencies now monitor the activities of their employees, sometimes covertly such as through video surveillance or checking electronic mail and files transmitted through servers. Filtering software may also be used to prevent staff from engaging in certain behaviours. 'Surfwatch', for example, can be customized to deny employees access to specified content. When the employee requests a site, the software matches the user's ID with the content allowable for the assigned category, then either loads the requested page or advises the user that the request has been denied. The software also logs denied requests for later inspection by management.

The use of computer software to monitor the business activities of government agencies also provides an effective means of detecting fraud and deterring people from acting illegally. The Australian Health Insurance Commission, for example, employs artificial neural networks to detect inappropriate claims made by health care providers and members of the public in respect of various government-funded health services and benefits. In 1997/98, this technology contributed to the HIC locating $7.6 million in benefits that were paid incorrectly to providers and the public (HIC 1998).

In addition, revenue authorities are able to make use of information derived from financial transaction-reporting requirements to identify suspicious patterns of cash transactions that could involve illegality or money laundering. In Australia, in 1997/98, the Australian Taxation Office attributed more than A$47 million in revenue assessed to its direct use of information provided by the Australian Transaction Reports and Analysis Centre (AUSTRAC). In one case, a taxpayer and associated entities had transferred more than A$1.3 million to a tax haven. Following an investigation, more than A$6 million in undeclared income was detected (AUSTRAC 1999).

Personal identification

Technology is able to offer considerable benefits in preventing fraud carried out through abuse of procedures for identifying people. In addition to the use of passwords or PINs, technologies of authentication based on space geodetic methods, biometrics and digital signatures are as applicable in this context as they are in the circumstances discussed in Chapter 2. Although biometric systems achieve much greater security than those that rely on passwords, they are expensive to introduce and raise potential problems in terms of privacy and confidentiality of the personal data stored on government computer networks. An initiative designed to reduce social security fraud in Toronto has been the enactment of legislation that would enable welfare benefit recipients to use fingerprint authentication when dealing with the Ontario government in Canada. Detailed privacy protection built into the legislation includes requirements for all biometric data to be encrypted and for the original biometric data to be destroyed after the encryption process has been completed (Cavoukian 1999).

Data-matching through the use of computerized databases is another way of validating identities and preventing fraud. In Australia in the early 1990s, a complex database was created by the Federal Government in an attempt to prevent taxation and social security fraud, by identifying people who have made claims for benefits from government funds to which they are not entitled. The Parallel Data-Matching Program makes use of tax file numbers and allows income records to be compared with payment records held by various benefit-providing departments. The program permits anomalies in payments to be identified and targeted for further investigation, and also enables individuals to be identified who are entitled to receive benefits they have not claimed. In the year 1996/97, the program resulted in direct savings of A$157 million for two departments, Social Security, and Employment, Education, Training and Youth Affairs. The cost of conducting the program for the same year was A$25 million, resulting in a net saving of A$132 million (Centrelink 1997).

The prevention of identity-related fraud is crucial when financial transactions are carried out with governments, and this will become increasingly important when agencies begin using electronic commerce. The Australian Government's *Project Gatekeeper*, for example, requires that individuals provide evidence of identification when registering with public key certification authorities (Australia, Office of Government Information Technology 1998). Certification authorities will require multiple and independent primary sources of identification. At present it is possible, when opening a bank account, to submit documents that have been forged by means of computerized desktop publishing equipment. This may also happen when people seek to obtain encryption key pairs from certification authorities.

Deterrence

The deterrent effects of criminal prosecution and punishment represent the final way of controlling electronic fraud against the government, though the future may see the resort to legal solutions decline in favour of technology-based, commercial solutions (see Lessig 1999). In addition to conventional judicial punishments such as fines and imprisonment, there may be various other consequences of the detection of fraudulent conduct, such as adverse publicity, professional disciplinary sanctions, civil action, injunctive orders and, most recently, various forms of reconciliation or community conferencing. Some of these may be particularly well suited to implementation in the digital environment. The confiscation of an offender's assets represents an effective means of deterrence as long as such sanctions receive wide publicity. Adverse publicity within government departments and forms of reintegrative shaming can also be effective in workplaces where reputations are important. One form of this that has been found effective in reducing unauthorized use of the Internet is employers publicizing details of websites visited by their staff and naming the staff member in question.

There are, however, a number of impediments to achieving deterrence through prosecution and punishment in the digital age (see Grabosky and Smith 1998). These concern the need to harmonize laws and procedures in order to accommodate cross-border conduct – a particular problem for those with federal systems of government in which activities regularly touch a variety of jurisdictions. Those who frame laws to deal with electronic forms of illegality are, however, approaching these concerns realistically by ensuring that laws are written in technology-neutral terms and are, as much as is possible, uniform (see Australia, Model Criminal Code Officers Committee 2000).

Returning to our pluralistic theme as described in Chapter 1, however, deterrent effects may also be achieved through the use of technology itself – an example of law being supported by code (Lessig 1999). One strategy developed to prevent software piracy by government employees, for example, entails the use of so-called Logic Bombs which are installed in programs. When activated through an act of unauthorized copying, the malicious code destroys the copied data and is even able to damage other software or hardware being used by the offender. Government employees who cause such damage would presumably be personally liable for replacement costs and any consequent loss. This use of technology might produce far greater deterrent effects than could be achieved through criminal action in the courts.

Conclusion

Computer technologies will greatly enhance the ability of people to defraud governments in the twenty-first century. Already a range of instances of such conduct have begun to emerge. Many security risks simply replicate traditional forms of public sector illegality, but make use of computers to carry them out faster and more efficiently. Others are directed at computer systems themselves, by theft of hardware and software or by using computers to transfer funds illegally.

A wide range of strategies exists to control and prevent such crime, some of which make use of well-established fraud control practices such as risk assessment and the provision of information to those most at risk, while others make use of the most recent digital technologies to prevent systems from being put to improper purposes or to detect illegal conduct immediately it takes place.

To reduce the risk of fraud against governments in the twenty-first century, it will be essential for all those involved to work cooperatively in making use of the latest technologies of computer crime control. The control of electronic fraud will, therefore, be a matter both for government agencies and for those in the private sector. Although public sector operations may be diminishing in scope, there remain abundant opportunities for people with the necessary expertise to compromise existing security procedures. Most likely, governments will need to call on the private sector to help in devising effective means of combating fraud in the years to come.

CHAPTER 5

Telephone Fraud and Theft of Internet Services

In 1963, the Quaker Oats Company began a national marketing campaign in the United States for a breakfast cereal known as 'Cap'n Crunch'. The product, aimed at a youthful market, was made of corn and oats, and its television commercials featured an eponymous seafaring character with a distinctive voice. It was not uncommon in the 1960s for the manufacturers of children's breakfast cereals to include small trinkets in the cereal box, to enhance the product's marketability. So it was that the manufacturers of 'Cap'n Crunch' included a small plastic whistle with their cereal product.

As it happened, the tone produced by the whistle packaged with the cereal corresponded closely to the 2600 cycle tone used in the United States long-distance telephone system. Although the significance of this was not apparent to either Quaker Oats or to the vast majority of consumers, it was not lost on one John Draper, who discovered that the whistle could be used to obtain telephone services without paying for them. In the rapidly growing underground movement whose members came to be known as 'phone phreakers' Draper became somewhat of a celebrity. He adopted the alias 'Cap'n Crunch', and is now a legend (Clough and Mungo 1992).

The challenge of defeating complex telephone systems has proved irresistible to many since the 1960s (Grabosky and Smith 1998). Over the last decade, as the number of telecommunications services available to the public has risen steadily, so too has the theft of such services. It has been estimated that cellular phone fraud alone in the United States costs some US$350 million a year, with the global cost of telecommunications fraud put at US$13 billion. Both these estimates are considerably higher than comparable figures from the mid-1990s (see Smith 1998: 65).

This increase in crime has been characterized by a shift towards 'process' or 'system' fraud as new protections have rendered old methods of fraud far more difficult to perpetrate. 'Boxing' into telephone switches has given way to fraudulent activity involving abuse of credit cards and identity-related fraud. The motivations of perpetrators remain varied: from the phreakers and hackers whose aim was largely to satisfy intellectual curiosity, through to organized criminals intent on making large profits. An additional incentive is the simple pursuit of prestige. Former hacker Peter Shipley, for example, has spoken of a desire for peer respect as the motivation of many hackers (Wired News 1998a). Most prosecutions to date have proved difficult because of the problems of establishing beyond reasonable doubt both the link between the accused and the crime, as well as criminal intention on the part of the accused.

The response of telecommunications carriers and Internet service providers has been to adopt cooperative risk management strategies and sophisticated investigative techniques in an attempt to discover and to minimize fraud and its costs, which range from uncollected revenue to indirect costs such as negative publicity and loss of goodwill. Such actions are often taken in preference to informing policing agencies.

This chapter will examine the latest developments in theft of telecommunications services; the changes that have taken place in the character of the crimes; the motives of the perpetrators; and the enforcement implications of recent developments in this area. (For an account of the position in the mid-1990s, see Smith 1998.)

Fraud and Telephony Services

Subscription fraud

The main form of telecommunications fraud that has taken place to date is subscription fraud (that is, the failure to pay for services used). Detection of such fraud is difficult because it is easily masked as bad debt. In subscription fraud, offenders typically misrepresent their identity in order to avoid payment. Misrepresentation of identity is also important because in the most severe of cases, subscription fraud is not an end in itself but rather a platform for several other frauds.

Call-sell fraud

One of the more notable frauds is 'call-sell fraud', which involves illegally onselling telecommunications services through either fixed-line or mobile connections. At present the most prevalent call-sell frauds involve people who have misrepresented their identity, gaining connection to a service at a residential address and setting up three-way call conferencing, which allows three different locations to participate in the same telephone conversation simultaneously. The fraud is perpetrated by having one person acting as an operator. This person dials the prearranged number, connecting the recipient of the call, and then connects their client, thus providing cut-price long-distance or international phone calls.

Such call-sell schemes often involve clients from particular ethnic groups. One way of conducting such a scheme is to solicit clients through advertisements in the client's language. The use of a foreign language serves the dual purpose of attracting clientele and avoiding police or other unwanted attention. Advertisements for clients to participate in such schemes state the cut-price rates and list a mobile telephone number which the clients are instructed to call in order to book their call. These advertisements are posted at venues that are frequented by members of the target nationality. Once the clients call the number, they are required to provide several details such as the telephone number that the illegal service provider is to call, and the expected duration of the call. The operator then informs the client of the cost of the call and arranges a time and place to meet and prepay the charges. The operator usually runs the scheme from premises rented under a false name.

Part of the attractiveness of this type of scheme for clients is that it enables people to gain access to international calls who might not otherwise have such access. The calls usually take place as agreed between the call-sell operator and the client, thus ensuring a loyal customer base. This loyalty is important, as the complicity of the client makes detection of the operation far more difficult for both telecommunications providers and policing agencies. Loyalty also provides a good customer base, which is vital as the call-sellers charge significantly less than the normal rate. With a loyal customer base and few overheads, profit from such fraud is usually large and can be virtually untraceable.

A variant on the call-sell operation involves capturing 'calling card' details and onselling calls charged to the calling card account. This relies to a degree on telecommunications carriers being too lax to notice a large volume of international calls being placed at limited localities atypical of the legitimate user's caller profile.

In March 1999, for example, several families in western Sydney were caught selling calls to Lebanon in this manner from their own homes (*Sydney Morning Herald*, 5 March 1999). The calling cards in this case were owned by citizens in the United States issued by several telecommunications providers also in the United States. The offenders advertised within the Lebanese community a mobile phone number through which interested parties could place bookings for international calls. The offender would take the number of the client and the family would then ring the client back. Eventually this fraud was discovered by an Australian telecommunications provider who found that the families in question had been placing abnormally large numbers of three-way conferencing calls from their homes. The demographics of the suburb and the locations of the conferenced calls also raised suspicions about the provider, leading to the discovery of the fraud. The telephone companies in the United States were unaware that their calling card numbers were being used for some calls that lasted for 80 hours continuously. Overall, it was estimated that the fraud cost the companies in the United States up to the equivalent of A$500 000 a month.

Call-sell fraud presents substantial difficulties of investigation and prosecution, as both the credit card and calling card frauds are often run by organized groups of offenders who may not even be from, or physically located in, the country of origin of the activity. If the operator is captured, the operator is usually under instruction to feign ignorance. In several cases investigated but not prosecuted in Australia, people claimed that they were 'doing a favour' for a friend they had just met, in return for lodging. Adducing evidence of intention on the part of the operator to actually engage in, or aid and abet such fraud is very difficult in the absence of a confession, and prosecution often fails through lack of evidence.

Where, however, offenders make use of 'switch boxes' (typically located at rented premises) in order to bill calls to a legitimate subscriber's account, prosecution is much more likely to succeed. Once the offender is identified, intention is fairly simple to demonstrate because of a preponderance of physical evidence and the technical nature of the guilty act.

Roaming fraud

Roaming fraud is similar to call-sell fraud in that it involves onselling mobile services at cheap rates. The main difference is that roaming fraud usually involves an offender who is located in another country. This type of fraud is difficult to prevent as it involves customers engaging an international or long-distance service having misrepresented their identity and then billing calls to the new account from interstate or overseas locations.

An example of this occurred in 1998 in Australia where a customer living in Western Australia took out a service (in conjunction with the purchase of a new mobile phone) in Victoria and persuaded the provider's customer service officer to lift the international bar late one Friday evening, thus allowing long-distance calls to be made (an international bar is a software protection which prevents users from being able to access their mobile service outside particular jurisdictions). Lifting the bar usually involves payment of a bond, as security for payment of the calls by the customer. In this case, a bond of A$200 was paid using a counterfeit credit card. In six days, the perpetrator had made 600 calls, mainly from Italy, valued at more than A$25 000. The perpetrator was able to be identified by the person who had sold him the service, eventually leading to arrest and prosecution. Because offenders are required to carry out a number of steps in order to perpetrate roaming fraud, prosecutions tend to be more successful; it is usually impossible to deny involvement and intention when a complicated sequence of events has taken place.

Prepaid mobile services

Each new product and service generates its own level of risk, which must be managed. Although products and services should, ideally, be assessed for risk before introduction and marketing, commercial pressure often causes strategic considerations to be overlooked. One instance where this has caused particular problems is in the area of prepaid mobile phones.

'Prepaid' mobile phones are available in most countries through various retail outlets. Their advantage is that the purchase transaction is swift, as there are no forms to be completed. If conducted in cash, the purchase is also effectively anonymous, thus presenting difficulties for law enforcement agencies and service providers alike. Fraud perpetrated through the use of prepaid mobile telephones would be difficult to prevent even in the presence of credit checks and identity checks. (See Smith [1999] for a discussion of the ease with which identities can be misrepresented for fraudulent purposes.)

Prepaid mobile telephones work by 'charging' each phone with a certain dollar value worth of calls. Like an arcade game, access ceases when the limit is reached. Some countries require the purchase of a card (the size and shape of a credit card), while in others the service can be recharged over the phone by authorizing a credit card account to be debited. New cards may, however, be purchased using stolen or altered credit cards. New Zealand Telecom, for example, adopts a system in which the phone can be recharged by credit card over the telephone. In order to minimize fraud, a limit of NZ$200 is placed on the purchase of extra time for each transaction. But this control was able to be circumvented by offenders using counterfeit credit cards for multiple transactions in quick succession. To diffuse suspicion, up to twenty other innocent people's prepaid accounts would receive the same amount, so that the use of the card could not easily be traced to the offender. The real concern surrounding prepaid telephones is not fraud per se but the fact that they effectively provide criminals with anonymous telephone services.

Dealer fraud

Telecommunications carriers are also at risk of fraud from their agents as well as at the hands of their clients. Dealer fraud, which can take several forms, is difficult to detect as it too can be masked as bad debt. One particular type that is difficult to detect is the creation of phantom customers through the use of fraudulent applications. Here the dealer makes up customers and 'sells' mobile phone services to them. This is a low-cost fraud to perpetrate, as some packages only require minimal upfront payments, with the bulk of the payment to be settled in the first billing period. The mobile phones are then able to be used for personal calls, or various call-sell schemes.

In October 1998, Vodaphone Australia terminated service to 40 'bad debt' customers, none of whom had made a single payment in more than 90 days, but who together had placed calls valued at more than A$20 000. It was assumed that at least some of the 'bad debt' represented dealer fraud, but this was difficult to establish. Sometimes the incentive systems established by carriers create new opportunities for fraud to occur. In one case, for example, a dealer used details of shelf companies to generate 100 corporate connections in order to gain A$30 000 in connection bonuses. Such crimes are typically opportunistic rather than organized, and not thoroughly planned, thus making it possible that more evidence survives.

Phreakers

Phreakers (telecommunications hackers, whose activities can be traced back to the 1960s: Smith 1998) are not a dying breed despite the introduction of technological measures such as encryption, which have made various forms of phreaking far more difficult to perpetrate.

In recent times, the Internet has allowed phreakers to develop communities of interest and to share their knowledge and experiences. FAQs (lists of frequently asked questions and answers) on phreaking have developed into an art form. Organizations such as the Australian League of Cypherpunks or Phone Losers of America (2000) provide a detailed history and theory of telephone networks in their locality and rudimentary comments on the best and worst locations from which to attempt various activities (Man Eats Dog 2000). Overall, the phreaker's objectives are to circumvent technical crime prevention measures. Follow-up activities vary from actually placing calls, through using the calls for harassment, to using the free services to mask hacking or more serious criminal activity.

Current popular targets for phreakers include carrier evaluation 'beta' products and services, chat/party lines and voicemail systems. In the United States in particular, voicemail systems are soft targets that are put to a variety of illegal uses, from audio bulletin boards to more sinister uses such as setting the voicemail to allow reverse charges, billing third parties and seeking calling card numbers. Where phreakers go to the trouble of engaging in such fraud, their purpose is usually to obscure computer hacking activity.

Fraud and Theft Control

Identification of fraud and preparation

The problem with many forms of telephone fraud is that fraud is sometimes masked by bad debt. Key tasks in preventing fraud of this nature involve conducting fraud vulnerability studies (particularly in respect of new services and surrounding key events such as Olympic Games), engaging in various revenue protection studies, and educating staff about fraud indicators (Hipkin 1999).

One vital element is cross-network cooperation. Theft of telecommunications services usually involves either credit card fraud or identity-related fraud, which must form the basis of risk management strategies. Recidivist offenders minimize their chances of detection and arrest by moving from carrier to carrier, despite adopting essentially the same modus operandi. In an attempt to minimize exposure to fraud, carriers have tried to adopt a model of cross-organizational cooperation, facilitated by the fact that in most countries elements of security and risk management are now considered to be non-competitive.

Anti-fraud systems

Fraud management systems also make use of real-time detection and reporting systems which enable fraud to be identified rapidly and losses minimized. Several advanced systems rely on the standardized protocol SS7, which acts by identifying variables such as the calling number, the number called, and the start and end times of the call. Because the systems rely on the SS7 variables, hacking into the systems to change billing details and similar activities is practically impossible, as each second is monitored and the result would be a readily identifiable gap in time. Such a protocol was used extensively to circumvent fraud at the 1996 Atlanta Olympic Games.

Anti-fraud systems also rely on previously identified 'red-flag' activities. Such activities include the use of three-way call conferencing facilities; long-distance access established by telephone followed by reverse call charges accepted from overseas; high-volume usage over short periods before disconnection; a large volume of calls where one call begins shortly after the termination of another; and the non-payment of bills.

The more sophisticated systems flag long-duration calls, and check high international direct dialling call usage as well as adopting customer 'fraud scoring'. National telephone carriers such as Telstra can compare 'normal' telephone usage to the usage of a customer and factoring that data into the customer's date of birth, occupation, location, credit details and other indicators to arrive at a percentage risk of fraud by that customer (see, for example, Telstra's 'Hunter' system).

Investigation

Owing to the highly technical nature of telecommunications fraud investigations, law enforcement agencies are often not involved until the closing stages of the investigation. At present investigation is typically carried out in-house, cooperatively with other carriers both nationally and internationally (Wallace 1999), or with the cooperation of security consultants. The first step for a carrier in an investigation is to determine the priority of investigations so that resources are concentrated on those investigations that are considered either to be vital or to have the best chance of success.

The object of most investigations is to prepare a brief of evidence which enunciates each element of the crime alleged and ensures that the evidence will be admissible and persuasive in court. Investigators have often come from law enforcement agencies and are aware of procedures for the collection and preservation of evidence. The task of the investigator involves a number of activities such as ensuring that facts are documented chronologically, preparing statements from potential witnesses and expert witnesses, and collecting company documents and records in order to connect these with witnesses for the purposes of identification. Investigators must also ensure that a chain of custody is transparent for all exhibits, that original records are preserved for forensic purposes, and that staff take contemporaneous notes.

Investigations are aided by policies which ensure that problems and incidents are reported to staff, regulators, the press and the public, so that all can cooperate in the investigation.

Enforcement

Once the investigation is complete, several enforcement options arise (Hipkin 1999): reporting the incident to the police and handing over all evidence for prosecution; immediate disconnection of the offending number or, alternatively, blacklisting (refusing further service) of the address. Other in-house options include closing the technical loophole that allowed the problem to occur, improving internal controls, or simply ignoring the incident.

One particularly clever method of enforcement sometimes adopted by Telstra, Australia's major carrier, involves billing the perpetrator for the full amount of the misappropriation rather than prosecuting the offence. This has the advantage of providing an immediate response in which the perpetrator is confronted with either disconnection (which may damage her credit rating), continuing requests for payment, or ultimately bankruptcy.

'Theft' of ISP Services and Time

Internet service providers are increasingly becoming the victims of theft or denial of connection services and time. This can take various forms, from hacking to obtaining services fraudulently, through to denial of service attacks. The last of these can crash ISP servers, costing them goodwill as well as the income they would ordinarily have earned had they remained operational.

By far the most common misappropriation of ISP time involves one account harbouring multiple users – usually in contravention of the supply contract. ISPs tend to perceive such individuals as potential customers rather than freeloaders, and so tolerate this behaviour. Similar misappropriation involves adhering to the letter of special offer agreements which charge for the number of times the client logs into the ISP rather than charging by volume or time. Such agreements often rely on a function that closes a connection if it has remained idle for more than a stated period. Some people have circumvented this by developing programs and devices that mimic being online so that they can remain online indefinitely, even in their own absence.

Ironically, hacking to gain free access is one of the less destructive actions that can be perpetrated against ISPs; it typically occurs on a small scale and involves a single user misappropriating ISP time. A more notorious incident, involving hacking on a grand scale, was the AOL4Free case – the first successful computer fraud prosecution in the United States involving an online network (Pelline 1997). In this case, a Yale University student, Nicholas Ryan, also known as 'Happy Hardcore', created, distributed and used his software 'AOL4Free' between June and December 1995 to allow himself and others to gain free access to America OnLine without paying the standard fee of US$2.95 per hour. AOL4Free worked by exploiting a weakness in AOL's billing system that enabled false commands to be sent to the AOL network to convince the billing system that the user was always using a free area of the service (Ryan 1997). When it was later discovered that this was detectable, a modified form of activity was used (Duva 1997).

Ryan received a sentence of two years' probation, six months of home confinement and a fine of US$50, and he was ordered to pay restitution of US$62 000 (Festa 1997). He commented in an editorial later that 'my crime is that of outsmarting you, something that you will never forgive me for'. Although not imposed in this case, it has become increasingly common for judges in the United States to impose a ban on computer use or Internet access as a condition of probation or parole for convicted hackers and fraudsters (Hesseldahl 1998).

Greater damage may be caused to ISPs' goodwill by attacks on the security of sensitive customer information. One incident involved an offender known as 'Phishers', who gained access to a database containing AOL users' passwords. He made use of their accounts in order to upload a Trojan horse (a program which appears to be useful or entertaining but which conceals a destructive purpose; a Trojan horse often destroys files or creates a 'back door' entry point that enables an intruder to access a system). Using this program, he was able to steal other AOL users' passwords. The stolen accounts were then used, causing the legitimate account-holder to be debited with charges incurred by the hacker (Fields 1997).

Another incident involving AOL was the release of its 'Guide Tool' on the Internet. This tool enabled hackers to roam AOL freely, gaining access to passwords, personal information and credit card information in addition to the power to add and to delete users' accounts. In September 1999, over 200 users of the ICQ ('I Seek You') instant messaging system reported their passwords stolen and their accounts taken over by unknown users. Another Trojan horse program, disguised as a JPEG digital photographic image attached to e-mail messages, sent the user's ICQ password to the sender. This is part of an ongoing history of poor security at Mirabilis, a subsidiary of AOL. A similar attack on OperaMail, a free web service operated by AOL, was perpetrated a month later with the theft of over 10 000 AOL passwords, this time with the victims being informed by e-mail that their accounts had been compromised (Oakes 1999).

Such actions suggest the motive to be more bravado than personal gain. An instance of this was the circulation of passwords to Telia (Sweden's largest phone company and ISP) as a grassroots political campaign designed to force the company to change its tariff structure to include flat rates for user access (Glave 1998). The theft involved stealing the password file, by exploiting security holes in the Unix 'qpopper' process of Telia's mainframe.

Most Western jurisdictions have offences making unlawful any interference with telecommunications networks by way of fraud, hacking or sabotage. In the United States, wire (the conventional term for telephonic and telegraphic) and related fraud is prohibited by 18 USC § 1343. This is supplemented by state legislation such as article 165 of the New York State Penal Law and section 13848 of the California Penal Code. In Australia, computer crimes are covered by Part VIA of the *Crimes Act* 1914 (Cwlth) as well as specific fraud offences in part 21 of the *Telecommunications Act* 1997 (Cwlth) and supplemented by state legislation such as Part 6 of the *Crimes Act* 1900 (NSW). Offences relating to telecommunications carriers, including telecommunications fraud, are proscribed by Part VIIB of the Crimes Act, with no state counterpart due to constitutional limitations.

Policing and enforcement of telephone carrier and ISP fraud, however, continues to be an area in which companies self-police rather than seek the aid of state criminal investigation officers. Telecommunications carriers and ISPs, by virtue of their specialist knowledge, have a distinct advantage, both in terms of efficiency and effectiveness at detecting intrusion and knowing where and how to gather forensic evidence of the commission of a crime. At present, police are not always informed of the commission of such offences, even when evidence of a crime has been gathered by organizations – the corporate equivalent of individual vigilantism. Carriers are effectively determining who the state should and should not prosecute, who will be punished, and what punishments will be imposed. Carriers may also wish to avoid police investigations which might uncover the presence of dubious activities of the company itself. Finally, large corporations may wish to avoid the publicity about weaknesses in their security system that may result from official investigations (Schwartau 1999).

Although carriers are correct in their assessment of police forces being undermanned and underprepared in this area, and are reluctant to incur the costs associated with taking legal action, the exclusion of police action creates a dangerous precedent which tends to support organized vigilantism. This further serves to erode public confidence in policing. From the perspective of the police, however, there is the problem of law enforcement agencies being underresourced in the area of computer-related crime. Police frustration is understandable, therefore, when carriers demand prompt investigation of 'trivial' offences such as the theft of $15 worth of ISP time, which can prevent them from dealing with more important investigations more worthy of their time and resources.

Conclusion

Demonstrably, the criminal law has been unable to offer much aid in the area of theft of telecommunications services, largely because it is not easy to get positive evidence of the identity of an alleged offender. Police forces are often underskilled and undermanned in relation to such crimes. It is important to remember, however, that such crimes do not exist in a vacuum, and the priority given to the economic loss occasioned by theft of telecommunications services must be adjusted to acknowledge the continuing need for the policing of violent crime.

In an attempt to mitigate damage to their own interests, telecommunications service providers set in place technological controls, which they hope can identify potential problem customers or provide evidence after a crime. Most of these controls entail crime prevention and detection rather than the collection of evidence and prosecution.

One of the ironies in the area of telecommunications and for ISPs is that typically the greater the number of controls, the less accessible the service is to the consumer. Telecommunications providers have to balance their desire for efficient and effective controls with their objective of ensuring that good customer relations are maintained. In economic terms, the two objectives need to be balanced. This is especially difficult because enhanced protections, which indirectly benefit consumers, are often met with hostility. When personal identification numbers were implemented, they were reviled by some on grounds such as that they represented the 'commodification' of the human being – reducing people to numbers rather than individual customers. Similarly, caller identification technology led to an outcry about individuals' privacy being eroded. Each time the level of protection and service is enhanced in a technological manner, negative as well as positive aspects emerge.

In the area of theft of telecommunications services, we see the difficulties that arise when organizations seek, within the bounds of the criminal law, to take the law into their own hands. Self-policing by telecommunications carriers, and other organizations with both technical expertise and an incentive to defend against fraud, may be more effective than the criminal justice system in dealing with theft of telephone and Internet services.

CHAPTER 6

Online Securities Fraud

Finance is the lifeblood of an economy. Businesses require capital in order to start up, and usually require additional resources to maintain or expand their activities. In some cases they may simply reinvest their profits. But expansion on a significant scale may require more than this. Thus businesses may also seek to borrow funds or to solicit investments in return for the investor's share of future profit. One of the basic means by which this latter strategy is pursued in industrial societies is for businesses to solicit investments from the public through the initial public offering of shares, and for subsequent buying and selling of shares by investors who expect the value of the shares to rise or fall. Securities markets are thus integral to a nation's economic system.

One of the fundamental strategies of governance in modern democratic systems is to allow capital markets to flourish and thereby drive economic growth. But the simple theory that crime follows opportunity applies no less to securities markets than to other areas of social life. Since the invention of the joint stock company in the seventeenth century, securities markets have provided golden opportunities for the unscrupulous, and numerous traps for the unwitting.

Governments therefore seek to create a climate favourable to legitimate entrepreneurs and investors, but one that discourages criminality. The fundamental goal is a market which is (and is seen to be) fair and efficient. This requires something of a balancing act. On the one hand, the rules of the game should be perceived as just and fairly enforced. On the other, the offer of and subsequent trading in securities should occur in a timely manner without buyers, sellers and intermediaries having to bear excessive regulatory burdens.

In this chapter we look at how the advent of digital technology has facilitated crime arising from trade in securities. After a brief historical overview, we discuss traditional forms of illegality to which securities markets may be vulnerable, and the regulatory apparatus that exists to ensure fairness and efficiency in markets. We then discuss the impact of digital technology on securities markets and the new potential for criminal exploitation which accompanies these technologies. Discussion then turns to available countermeasures against securities-related crime using high technology. We first explore the adequacy of existing legislation and regulatory arrangements, then discuss the potential for self-regulation by the various institutions involved in securities transactions. We conclude by speculating on what the landscape of the securities industry, and its vulnerability to criminal exploitation, might look like in the years ahead.

There are markets, and there are markets. Descriptions of the early days of Change Alley in the City of London, or of the New York Stock Exchange, describe how brokers would meet in a coffee house, and how the president called each security one by one, with members wishing to sell indicating the availability of a security, and those wishing to buy, bidding for them. A secretary would transcribe the prices at which sales were concluded and these were published in newspapers. In 1820, 28 stocks were listed on the New York Stock Exchange (NYSE), and the average daily trading volume was 156 shares (Banner 1998a: 117). By contrast, on 7 April 1999 (to pick a more or less typical day) the NYSE consolidated volume was 997 525 530, with a total of 3566 consolidated issues traded.

In Australia, the rather unique nature of its establishment and early history as a penal colony delayed the formation of a financial system. It was not until 1828 when a Matthew Gregson was given permission to deal in Bank of New South Wales shares. By 1837, the first Australian stock market was established, and the *Sydney Morning Herald* commenced daily publication of price quotations. The Melbourne Stock Exchange was established at the end of the 1850s as a result of the economic activity inspired by the gold rush. Each Australian capital city had its own exchange until the formation of a national exchange, the Australian Stock Exchange (ASX), in 1991.

Traditional Forms of Sharemarket Illegality

Although there are many permutations of illegality relating to securities markets, they take three basic forms: misrepresenting the underlying value of a security at the time of the initial public offering; market manipulation during secondary trading of a security; and insider trading. There are, in addition, various other possible offences that relate to irregularities in mergers and acquisitions and to breaches of fiduciary duties by professional advisers. Examples of these are embezzlement of client funds and a broker's gratuitous trading on a client's account in order to generate commissions – known in some jurisdictions as 'churning'.

Misrepresentation

Perhaps the classic offence relating to securities is the offering for sale of securities that are either essentially worthless, or whose underlying value is significantly overstated. A variety of individuals may benefit from misrepresenting a company's investment potential. Misrepresentations can be made by principals of the company whose shares are being offered or traded, by intermediaries such as underwriters and brokers, and by investment advisers. At the most gross level, they promise guaranteed returns. Unfortunately, the returns flow to the deceiver, not the deceived.

Misrepresentation is an elastic concept. At one extreme it may involve deliberate falsification with unabashed fraudulent intent. At the other it may involve naive ebullience and unrestrained optimism. The threshold of criminality may not always be clear.

The potential for misrepresentation of all kinds increases when speculative activity increases, and an atmosphere of general optimism prevails. An historical example of wild speculation in Australia's relatively recent history is the Poseidon nickel boom of 1969, when the price of shares in one mining company rose from 75 cents to A$280 in four months, and any company purporting to have anything to do with nickel attracted investor attention. In such an environment it is not difficult to imagine how ordinary citizens, lured by the prospect of instant wealth, may be seduced by offers too good to be true.

Myths and conjectures can be disseminated or amplified by irresponsible news coverage in the media. Simply riding a speculative bandwagon in an atmosphere of infectious optimism is bad enough. But reporting on financial matters has always provided opportunities for conflicts of interest. In Australia during the freewheeling days of the 1980s, it was not uncommon for financial journalists to be offered shares or options in newly formed companies. Inevitably there was a conscious temptation, or subconscious inclination, to accord the company favourable treatment in subsequent media coverage (Grabosky and Wilson 1989: Ch. 7).

Misrepresentation can also occur once trading in a security begins, such as when a broker or investment adviser recommends the purchase of a stock knowing it to be overvalued, or suggesting a sale of a stock with good growth prospects.

Market manipulation

The second basic form of illegality involving securities markets is market manipulation (Goldwasser 1998). Once trading in a stock commences, an offender will try to influence the price of a security in a manner enabling him to benefit from the price change. This may be accomplished by words as well as by deeds. The practice is as old as the sharemarket itself. Robb (1992: 11) relates how a seventeenth-century director of the Bank of England manipulated the price of government securities issued by the Bank for his own advantage. The annals of business are littered with examples of 'announcements' or rumours of impending discoveries which turned out to have been without foundation. These days, merely to encourage speculation that a company is close to developing an AIDS vaccine, or is discovering an ore body, may trigger speculative investment.

Of course, what goes up sometimes comes down. Not all misrepresentation is positive, and there are those who may seek to profit from negative fabrication. The use of rumour, hyperbole, or other forms of misinformation to boost the price of a stock before the manipulator's quick and profitable exit, is termed 'pump and dump'. The converse, where a manipulator talks down a stock so that he may buy in at a bargain price, is called 'slur and slurp' (Goldwasser 1998: 520).

People can also manipulate the price of a stock by engineering a pattern of transactions to attract the attention of the unwitting investor. For example, a person who holds a number of shares in a company may arrange to make an inflated offer for an additional small parcel in order to drive up the price of a stock. When the price rises, the person will then sell her shares. A related practice is the engineering of transactions that achieve a quick movement in the share price immediately before the close of trading. The artificially high bid may be withdrawn at the opening of trading, but the market meanwhile may be misled by the unusual activity.

Alternatively, the manipulator may place numerous buy and sell orders to create the illusion of large turnover, in order to attract the interest of unwitting investors. This can also be done by two or more accomplices acting in concert, a practice known as 'pass the parcel'. As is the case with verbal strategies of manipulation, those in a position to profit from a fall in the price of a stock may try to drive its price down in order to buy in at a bargain price.

Market manipulation is easier to achieve with what are called 'thinly traded' shares. These are shares in fairly small companies with relatively few shareholders where trading tends to occur in low volume, such as companies involved in gold exploration. By contrast, it would be extremely difficult for someone to manipulate the price of shares in General Motors or IBM.

Insider trading

The third basic form of illegality is known as 'insider trading', which occurs when a person acts on information from within a company, or knowledge not otherwise publicly available (Tomasic 1991). For example, a geologist working for a mining company who discovers a spectacular ore body is prohibited from purchasing shares in the company before a public announcement of the find. Similarly, a clerk in a company's accounting department who becomes aware of an impending financial report containing a great deal of red ink may not unload his or her shares before the report is released to the market. In some jurisdictions, the prohibition on insider trading will extend to 'outsiders' who may be privy to inside information, such as printers who are producing documents before their disclosure, financial journalists before the publication of previously confidential data, or employees of a law firm who are aware of an impending takeover announcement (Ziegelaar 1998).

To safeguard against insider trading, most securities regulators require that corporate insiders disclose any ownership and transaction of their company's securities.

These three basic forms of securities fraud are not mutually exclusive, and may occur in combination. For example, a company insider who is aware of looming financial disaster may arrange to have the prospects of the company portrayed in an overly optimistic light. She may also engage in trading to create the illusion of market confidence in the company's shares, then unload her holdings before the grim realities are disclosed and the price collapses.

Cynics would argue that there are a thousand ways to separate a fool from his money, so one should not be too paternalistic. But there are very good reasons to be concerned about crime relating to securities markets. The reason why these various forms of misconduct are made criminal is that if left unrestrained, they would place the ordinary investor at a considerable disadvantage, to the extent that he might shun the stock market as a place to invest. This in turn will make it more difficult for businesses to raise capital, with consequent adverse impact on the economy. Reputations for questionable practices may turn off more than the ordinary investor – large institutional investors may be similarly deterred and a process of capital flight may ensue. By way of illustration, after what became known in Australia as the corporate excesses of the 1980s (Sykes 1994), Australian shares became less attractive to investors both domestic and international.

Conventional Regulatory Responses

Information is the lubricant of markets. The essence of a fair market is transparency and symmetry of information. In the ideal market, access to information should be equal in quality and timeliness. Indeed, as Braithwaite and Drahos (2000) argue, the concept of transparency is increasing in importance in the regulatory debate surrounding the globalization of business.

The foundation for securities regulation is disclosure. The basic principle is that all information relevant to the value of a company should be available to the market. In order to protect against various forms of fraud, most nations with significant securities markets have developed strict regulations for offering securities. These entail a requirement that securities offered for sale be registered with the relevant regulatory agency, such as the Financial Services Authority (formerly the Securities and Investments Board) in the United Kingdom (FSA), the Securities and Exchange Commission (SEC) in the United States, or the Australian Securities and Investment Commission (ASIC) in Australia, and that the offer be accompanied by a prospectus, setting out in considerable detail the circumstances of the offer. Required disclosure serves to inform markets, which themselves may provide a disciplinary function, rewarding good performance and punishing the inferior.

In addition to setting out the required contents of a prospectus in minute detail (including the size of type; see Australia's *Corporations Law,* section 1021(1)(2)), in Australia the *Corporations Law* specifies a more general standard of disclosure. This requires that a prospectus contain:

> [all] such information as investors and their professional advisers would reasonably require, and reasonably expect to find in the prospectus, for the purpose of making an informed assessment of:
>
> (a) the assets and liabilities, financial position, profits and losses, and prospects of the corporation; and
>
> (b) the rights attaching to the securities (section 1022(1)).

Securities regulators further require publicly traded companies to report regularly regarding their financial condition and business operations (see, for example, the Australian *Corporations Law,* Part 2M(3)). In addition to regular reports, securities regulators may also require the interim disclosure of any developments that materially affect the operations of a company.

Beyond the direct regulatory activity of government agencies and the discipline of informed markets, a significant element of securities regulation has been the exercise of quasi-regulatory responsibility by third parties who themselves have been licensed by regulatory authorities. Thus an initial offering of shares usually requires the assistance of underwriters or investment bankers, who manage the offering in return for a fee. Subsequent share transactions are handled by brokers, who receive a commission on each trade. Licensing requirements may even extend to people providing investment advice but not necessarily acting as an underwriter, or broker/dealer. The licensing function is designed to ensure that participants in securities markets are people of integrity, and that the unscrupulous are excluded.

The licensed professional lends her integrity to the transaction, implicitly pledging her own reputation (Coffee 1997: 1203). The public image and the financial success of the underwriter will be tied to the integrity and the ultimate success of the offering. Brokers, too, are in a position to act as gatekeepers, since they may be in a position to detect manipulative practices and may safeguard clients from unwise investments.

Although in general underwriters and brokers can be part of the solution, some are part of the problem. In between the state and these third-party gatekeepers are other organizations, such as stock exchanges and associations of securities dealers, which themselves may play a quasi-regulatory role. In the United States, the National Association of Securities Dealers (NASD) may impose fines or suspend individuals and firms for violations of securities laws and NASD rules. Such decisions may be appealed against to the SEC and subsequently to the federal courts in the United States (Hyatt 1993). Stock exchanges may also engage in market surveillance. They may take disciplinary action against members, and may suspend trading in shares of a particular company when suspicious trading patterns become apparent.

This layered web of regulation, involving state activity, self-regulation, and an informed market, is the basic structure of securities regulation in advanced industrial societies today.

New Technologies and Their Impact

The convergence of computing and communications technologies in the digital age has led to the transformation of the securities industry globally. The kind of trading volume that occurs on a typical day in London or New York would simply not be possible but for electronic trading systems. At least one company has already held its annual meeting on the Internet (Prentice 1998: 2), and the New Zealand Stock Exchange has introduced mobile direct share trading with the use of Wireless Application Protocol (WAP) mobile phones linked through the Internet to the Stock Exchange that permits online access to the Exchange's trading system (Barker 2000).

This is not the first time that the securities industry has been transformed by technology. The advent of the telegraph, and later the telephone, greatly facilitated the exchange of information and enhanced the speed with which transactions could be made. Indeed, electronic trading has replaced the open outcry that traditionally characterized trading floors.

The digital age, however, appears to be unprecedented in the volume of information about investments currently available to ordinary investors, and the speed with which it may be obtained. And most of this information is free. No longer must investors telephone brokers to discern current price information, nor need they wait until the daily financial newspaper hits the streets, or the weekly investment newsletter to which they subscribe arrives in the post. Indeed, the traditional business hours of trading are becoming of less importance because investors in one country are able to trade through the night in another country whose stock exchange remains open (Preiss 1998).

Small investors, sitting at home, now enjoy immediate access to vast amounts of information, some more reliable than others. Conventional search engines permit individuals to search the web for information about a particular company. There has been a proliferation of online information services providing real-time stock quotations, various analytical data such as historical statistics on price and volume and price:earnings ratios. Brokers and investment advisers no longer have a quasi-monopoly on timely market information or in-depth research.

Even more revolutionary is the ability of investors to communicate with each other. New forums for investor interaction have developed overnight. The Internet abounds with chat rooms and discussion groups devoted to a wide variety of investment-related issues. Online bulletin boards enable small investors to communicate with fellow shareholders, and with various participants in securities markets, both amateur and professional. New technologies, therefore, have great potential to reduce the inequalities in access to information that previously impeded the ideal functioning of markets (Coffee 1997: 1196).

The digital age has also brought about much more efficient means by which companies may communicate information to investors and financial markets. Contrast the instantaneous widespread dissemination of information by digital means with the slow and costly use of paper-based media, with its uneven accessibility.

Inventors and entrepreneurs are now able to reach millions of potential investors without incurring the expense of printing, 'road shows', public relations consultants, and other expensive means of communicating with a large audience. For large and small businesses alike, the Internet provides a medium for much more efficient communication, and the advent of the World Wide Web enables companies to post announcements instantaneously.

One of the most significant developments is that the Internet has become a medium for the public offering of securities. This promises greater access to capital markets by small business, with the cost of capital formation being significantly reduced. In the United States, Spring Street Brewing Company completed the first online public offering in April 1996 (Coffee 1997: 1202). In Australia, the first Internet offering was made in July 1996, and subsequent offerings of managed funds products have also been made online (Simpson 1997: 372).

So too has the monopoly on transactions previously enjoyed by intermediaries begun to give way. Online systems have begun to enable individual investors to bypass intermediaries, who may have a vested interest in certain transactions (Coffee 1997: 1197), and subscribe directly to a share offer, or to engage directly in the subsequent purchase and sale of shares. A variety of ISPs and websites now offer direct links to brokers or dealers, and some traditional brokerage firms offer online services themselves. Services exist that allow individuals to execute, buy and sell orders online and in real time, with very inexpensive commissions, often as little as US$6 per transaction (Gleick 1998; Salmons 2000). Transactions can thus occur with unprecedented efficiency. The term 'day trading' has been coined to refer to the practice of online trading by private investors to take advantage of small changes in share prices. Millions of such accounts currently exist, and as is the case with the uptake of other forms of digital technology, a geometric growth rate is likely (Whitford and Rao 1998).

For all their benefits, however, new technologies pose significant risks. Some of the following traps have accompanied the advent of cyberspace investing. In general, they entail old frauds perpetrated with new media.

Securities Fraud in the Digital Age

Fraudulent offerings

Cyberspace now abounds with a wide variety of investment opportunities, from traditional securities such as stocks and bonds, to more exotic opportunities such as coconut farming, the sale and leaseback of ATMs, and worldwide telephone lotteries (Cella and Stark 1997: 837–44). But new technologies facilitate solicitation by illegitimate as well as legitimate offerors. Classic pyramid schemes and 'Exciting, Low-Risk Investment Opportunities' are not uncommon. In 1998, a worldwide clean-up operation, involving the Office of Fair Trading in Britain and its counterparts in 22 other countries, identified 1159 potential 'get rich quick' schemes being advertised on Internet sites (Office of Fair Trading 1998). In one Ponzi scheme, recently closed down by the SEC, franchises for ATMs were offered for a US$25 000 investment with a 100 per cent return. Approximately 200 machines were sold to 141 investors, with returns being paid to some. The moneys paid to the early investors were, however, taken from later investors rather than from profits, thus preventing later investors from reaping any rewards from their investments (Hogan 1997: 174–5).

The introduction of the Internet has been accompanied by unprecedented opportunities for misinformation. Fraudsters now enjoy direct access to millions of prospective victims around the world, instantaneously and at minimal cost. Those who use the Internet, however, have great difficulty in identifying those with whom they are communicating. Some organizations may intentionally disguise their identity through the use of remailing facilities in order to defraud individuals later and avoid detection. Others may simply be neglectful in providing accurate and verifiable information.

Because the technology of the Internet makes it easy to disguise one's identity (see Chapter 7), it is ideally suited to investment solicitations. In the words of two SEC staff:

> At very little cost, and from the privacy of a basement office or living room, the fraudster can produce a home page that looks better and more sophisticated than that of a Fortune 500 company ... Not just the plain vanilla home page of yesteryear, though; rather, investors are presented with sleekly designed, multi-page Web sites that include all the latest bells and whistles such as research engines, interactive graphics, sounds, video, and futuristic looking formats and links. (Cella and Stark 1997: 821–2)

One is reminded of the now famous cartoon that appeared in the *New Yorker* (5 July 1993: 61) featuring a canine computer user, bearing the caption: 'on the Internet, nobody knows you're a dog'. The same could be said about some investment solicitations. One of the more prominent early cases of irregularities in online investment solicitations concerned a proposal to finance the commercial acquisition and husbandry of eels. Daniel Odulo, an enterprising 19-year-old, posted a solicitation to two Internet news groups. It did not reveal that the entrepreneur lacked experience in aquaculture. Moreover, the proponent falsely stated that potential investors would be insured against potential losses, and the solicitation included fabricated endorsements from fictitious persons (*SEC* v *Daniel Odulo*, No. 95-424; see Robertson 1996: 402–3; Cella and Stark 1997: 838–9). The impecunious Odulo later agreed to discontinue his activities.

Another entrepreneur, who did not disclose his lengthy criminal record, advertised 'promissory notes' with guaranteed annual returns between 12 and 22 per cent in at least 21 Internet newsgroups (Fontana 1998: 312; Prentice 1998: 24). A direct public offering over the Internet for the company Interactive Products and Services raised US$190 000 from 150 investors. The principal pocketed the proceeds. He was the subject of civil action by the SEC, and state criminal charges in California, resulting in a ten-year prison sentence (SEC 1998). The legitimacy or otherwise of some investment information posted on the Internet may, however, be readily apparent. Investors who read of the discovery of the world's largest gold deposit, or who visited the 'Motley Fool Website' should clearly have been wary of any information disclosed (Haines and Johnstone 1998).

Sharemarket manipulation

The possibility of sharemarkets being manipulated through the use of new technologies is not new. In 1867, for example, a Wall Street stock broker collaborated with Western Union telegraph operators to counterfeit messages which reported bankruptcies and other financial disasters supposedly befalling companies whose stock was traded on the NYSE. When the share prices were driven down, the perpetrators then purchased their victims' stock (O'Toole 1978: 97). Similarly, the advent of 'ticker tape' invited attempts at manipulating price quotations (see Robb 1992: 90).

The accessibility of online trading facilities, however, has posed unprecedented opportunities for sharemarket manipulation. The proliferation of day traders contributes to the volatility of share prices, particularly in those securities that are thinly traded. Against this background, structuring transactions in a manner that will give the impression of momentum in the price of a share would appear to be readily achievable at the hands of an individual or investors acting in concert.

Online misrepresentation may be less direct, and more insidious. False information may spread unwittingly but widely through chat rooms; discussion groups contain vast amounts of information. Obscure, thinly traded securities may be the subject of attention by promoters trying to drive prices up and short sellers who seek to drive prices down (Eaton 1996: D1). Conflicts of interest on the part of those who are talking a stock up or down may not be apparent, and the use of multiple accounts by one individual, facilitated by technologies of anonymity and pseudonymity (Denning 1999: 313–16), may create the appearance of collective action. Indeed, when undertaken intentionally, such activity can be carefully orchestrated to make it appear that many people in different locations around a country are of like mind.

Rumours based more on aspiration than fact may circulate widely, and given the ease of direct access to online trading, information about a company can within minutes have an impact on the price of a stock. In the United

States, the suggestion that one company's fingerprint identification technology would be adopted by a major credit card company led the share price to increase almost overnight from US$0.03 to US$1.75. The rumours proved to be unfounded and the company, Comparator Systems Corporation, was wound up not long afterwards (Coffee 1997: 1224). In another case, America OnLine and Prodigy Bulletin Boards carried stories of a major diamond discovery by a small Canadian company. The price increased sixfold in less than a month from Can$0.30 to Can$1.85 (Robertson 1996: 408).

It appears that unsolicited hot tips ('spamming' – see page 119) are becoming more common. Bulk e-mail programs allow stock promoters to send personalized messages to thousands and even millions of Internet users simultaneously. In a recent Australian prosecution, two people were alleged to have sent more than four million e-mail messages in May 1999 to recipients around the globe containing false and misleading information relating to the likely increase in the stock price of an American company. The messages contained a statement that share values in the company would increase from the then current price of US$0.33 to US$3.00 once pending patents were released by the company, and that the price would increase up to 900 per cent within the next few months. Similar information was posted on the Internet. As a result the company's share price doubled on the NASDAQ, with trading volume increasing by more than ten times the previous month's figures. ASIC prosecuted the offenders for distributing false and misleading information with the intention of inducing investors to purchase the company's stock. In October 2000, one of them was sentenced to two years' imprisonment (21 months of which were suspended) in the County Court of Victoria (Lindsay 2000). Unsolicited investment promotions can be indiscriminate – in September 1998 the chairman of ASIC received an anonymous e-mail message enthusiastically promoting a stock traded in the United States (Phillips 1999: 13).

While companies and their principals may be the source of misrepresentation, they are often the unwitting subjects. Home pages on the World Wide Web are not always secure. In 1997 access was gained to the web pages of both the US Department of Justice and the CIA and the pages defaced by hackers (Denning 1999: 228). During the 1998 Australian federal election the web page of the Liberal Party of Australia was similarly altered. Corporate home pages are no less vulnerable, and can be altered less obtrusively with market-sensitive data (Prentice 1998: 47). In one case, a disgruntled website manager placed disparaging information on his company's page. It may not have been entirely inappropriate, as there was in fact fraud by the company (Prentice 1998: 49).

In the digital age, misinformation, and rumours derived from it, can spread very fast. One could imagine a situation in which a hacker or manipulator would buy shares in a company, then make subtle changes to the company's web page which would allow him to benefit from resulting changes in the company's share price before the hack was detected and rectified (Prentice 1998).

The use of counterfeit web pages for purposes of misinformation has already been documented. These can convey an illusion of legitimacy, complete with links to home pages of regulatory authorities and other institutions. Websites such as www.angelfire.com exist which enable users of the Internet to create their own home pages. In April 1999, a fabricated page resembling a product of the financial news service Bloomberg posted a report that a particular company was about to become the subject of a takeover. The report included apparently credible quotations, purportedly from officials of the company. Within minutes, the stock was being touted in chat rooms, and the price rose by more than 30 per cent before the hoax was discovered (Associated Press 1999a).

As has been the case with corrupt financial journalists and other third parties, publishers of investment newsletters may be co-opted for the purpose of misleading the investing public. An electronic newsletter, *SGA Goldstar*, posted information encouraging the purchase of stock in a company called Systems of Excellence Inc. (SEXI). In return, principals of the newsletter received 300 000 shares of SEXI stock (Fontana 1998: 313). One firm sent unsolicited e-mail to 500 000 potential investors recommending stocks, and were paid for their efforts. Another firm recommended a number of stocks on its website in return for cash and stock options, but failed to disclose that they were being compensated for their services (*New York Times on the Web*, 26 February 1999).

The bull market of the 1990s, combined with the advent of online trading by ordinary investors, has made for some disappointment. Delays in the execution of trades may see significant price changes occurring between the time a bid is made and the time a trade is executed. Those investors who fail to set a ceiling on their bid to purchase may wind up paying much more for a stock than they originally intended. The unwary and the incompetent may be strikingly unsuccessful investors, at substantial personal cost. In July, 1999, for example, Mark Barton, an unsuccessful day trader, killed his wife, children, and a number of other people in Atlanta, Georgia. Regulators in the United States have received many complaints from unsophisticated day traders who have paid high commissions for the wrong investments. Not all of this, however, is criminal activity.

Further along the continuum of criminality, it may be possible to hack into an online trading system itself, and reroute orders, although no cases of this have yet become apparent.

Digital insider trading

The possibility of insider trading also exists in the digital world of sharetrading in much the same way as in the traditional sharemarket. Already instances have begun to emerge. In Australia, beginning in 1999, a number of small speculative mining companies began to diversify into Internet-related activities, hoping to ride the wave of the high-technology wave on the

Australian and United States stockmarkets. Regulatory authorities observed a surge in stock price and trading volume of these companies *before* the announcement of their proposed metamorphosis, and expressed concern about the future potential for false and misleading statements in this area (Phillips 1999).

Other opportunities for dishonesty

The advent of online share trading may create opportunities for other types of dishonest conduct, most obviously related to electronic funds transfer fraud. But since traders are usually required to maintain an account from which funds are drawn to pay for a transaction, this would appear less of a problem. So too would electronic impersonation, and the fraudulent conversion of funds to another person's account.

More problematic, however, is the use of the Internet to sell investor education materials that in some cases may be of little use or even counterproductive. One of the more recent instances of this relates to deceptive day trading promotions. In the United States, the Federal Trade Commission (FTC 2000) has recently targeted Internet operators who sell combinations of online 'real-time' training, software programs, trading manuals, e-mail newsletters, and mentoring services for prices ranging from US$79 to US$4995. Some services have claimed average returns of up to 167 per cent and included deceptive testimonials and little or no mention of the risks associated with day trading. The FTC has taken action against a number of companies promoting such advertisements. In the United Kingdom, the Financial Services Authority (FSA 2000) has also issued warnings concerning the risks associated with day trading.

Countermeasures

As is the case with many other domains of cyberspace illegality, the control of offences relating to securities markets will require various countermeasures (Grabosky and Smith 1997). These will involve a combination of state regulatory controls, self-regulation by individual companies and industry associations, and quasi-regulatory activity by third parties. Although the legal response will remain important for some time to come, the technological, commerce-based response described by Lessig (1999) as regulation by code may well take on primary importance in the digital age.

The legal framework

Most forms of securities fraud that involve the use of digital technologies can be addressed by existing laws in most Western industrial nations. In the United Kingdom, fraud in relation to investments can result in proceedings being taken for a range of offences including obtaining property by deception, evasion of liability by deception, false accounting, publication of misleading statements about a company's affairs, conspiracy to defraud, and even theft if investment funds are stolen (Arlidge, Parry and Gatt 1996: 293). In addition, the *Financial Services Act* 1986 (Eng.) creates various offences relating to false statements, market manipulation and insider dealing. Section 47(1) of the Act provides that:

> Any person who –
>
> (a) makes a statement, promise or forecast which he knows to be misleading, false or deceptive or dishonestly conceals any material facts; or
>
> (b) recklessly makes (dishonestly or otherwise) a statement, promise or forecast which is misleading, false or deceptive,
>
> is guilty of an offence if he makes the statement, promise or forecast or conceals the facts for the purpose of inducing, or is reckless as to whether it may induce, another person (whether or not the person to whom the statement, promise or forecast is made or from whom the facts are concealed) to enter or offer to enter into, or to refrain from entering or offering to enter into, an investment agreement or to exercise, or refrain from exercising, any rights conferred by an investment.

Section 47(2) also creates an offence of market manipulation. It provides:

> Any person who does any act or engages in any course of conduct which creates a false or misleading impression as to the market in or the price or value of any investments is guilty of an offence if he does so for the purpose of creating that impression and of thereby inducing another person to acquire, dispose of, subscribe for or underwrite those investments or to refrain from doing so or to exercise, or refrain from exercising, any rights conferred by those investments.

Both these offences carry maximum penalties (on indictment) of up to seven years' imprisonment or a fine or both, or (on summary conviction) six months' imprisonment or a fine (s. 47(6)).

Insider dealing may result in infringement of section 47(1) of this Act by dishonest concealment of material facts, or infringement of the specific prohibitions against insider dealing contained in Part V of the *Criminal Justice Act* 1993 (Eng.) (ss. 52–64 and schedules 1 and 2). Each of these offences generally requires that the conduct or the victim be located within the United Kingdom in order to satisfy the jurisdictional requirements (ss. 47(4) and (5) Financial Services Act and s. 62(1) Criminal Justice Act.

In a release issued on 28 May 1998, the FSA provided substantial guidance on the treatment of investment material on Internet sites accessible in the United Kingdom but not intended for investors in that country. Where people in the United Kingdom are able to read material on their computer screens, this will be taken as having been issued in the United Kingdom and accordingly must comply with the provisions of the Financial Services Act.

The FSA examines individual cases separately and considers the following factors in deciding whether or not to prosecute:

- whether the site is located on a server outside the United Kingdom;
- the extent to which positive steps are taken to prevent access to the site from the United Kingdom;
- the effectiveness of a firm's system for ensuring that only persons who may lawfully receive the investment services or investments do so; and
- the extent to which any advertisement is directed at persons in the United Kingdom (see MacHarg and Clark 1999: 475).

Some of the matters that the FSA takes into account in deciding whether an Internet-based investment is directed at people in the United Kingdom include whether the site has links to other sites which prohibit access by people in the United Kingdom; whether the site has been notified to those who maintain search engines as a United Kingdom site; whether associated bulletin boards or chat rooms have been used to promote investments in the United Kingdom; and whether the site has been advertised in other media in the United Kingdom.

In the United States, the three federal statutes applicable to securities fraud in its traditional or digital manifestations are the *Securities Act* of 1933 which governs the issue of securities by companies, the *Securities Exchange Act* of 1934 which governs the trading, purchase, and sale of those securities, and the Investment Advisers Act of 1940 which regulates the activities of investment advisers. They and their subordinate regulations are broad enough to embrace the new media by which frauds may be committed. For example, section 17(a) of the Securities Act of 1933 provides that:

> It shall be unlawful for any person in the offer or sale of any securities by the use of any means or instruments of transportation or communication in interstate commerce or by the use of the mails, directly or indirectly
>
> (1) to employ any device, scheme, or artifice to defraud, or
>
> (2) to obtain money or property by means of any untrue statement of a material fact or any omission to state a material fact necessary in order to make the statements made, in the light of the circumstances under which they were made, not misleading, or
>
> (3) to engage in any transaction, practice, or course of business which operates or would operate as a fraud or deceit upon the purchaser.

Section 9(a)(2) of the Securities Exchange Act of 1934 also makes it unlawful:

> for any person, directly or indirectly, by the use of the mails or any means or instrumentality of interstate commerce, or of any facility of any national securities exchange, or for any member of a national securities exchange ...

(2) To effect, alone or with one or more other persons, a series of transactions in any security registered on a national securities exchange creating actual or apparent active trading in such security, or raising or depressing the price of such security, for the purpose of inducing the purchase or sale of such security by others.

The SEC has made it clear that these provisions are fully applicable to electronic communications no less than to communications on paper (Prentice 1998: 4). In addition, section 10(b) of the Securities Exchange Act of 1934 provides:

> It shall be unlawful for any person, directly or indirectly, by the use of any means or instrumentality of interstate commerce or of the mails, or of any facility of any national securities exchange …
>
> (b) To use or employ, in connection with the purchase or sale of any security registered on a national securities exchange or any security not so registered, any manipulative or deceptive device or contrivance in contravention of such rules and regulations as the Commission may prescribe as necessary or appropriate in the public interest or for the protection of investors.

The Investment Advisers Act of 1940 also makes it unlawful (*inter alia*):

> for any investment adviser by use of the mails or any means or instrumentality of interstate commerce, directly or indirectly
>
> (1) to employ any device, scheme, or artifice to defraud any client or prospective client;
>
> (2) to engage in any transaction, practice, or course of business which operates as a fraud or deceit upon any client or prospective client …

In addition to these federal legislative measures in the United States, several states also have an enforcement role under their own laws. Generally, state-based laws focus on individual investor protection issues, while national agencies deal with matters of broad-based market concern.

The SEC also has broad rule-making powers which can be applied to online share trading. Its response to the advent of online public offerings has been cautiously permissive. It might best be characterized as incremental flexible regulation. Rather than articulate fixed policies, the SEC reviews proposals case by case, issuing 'no action' letters when a given offering is condoned. In some cases where there has not been much in the way of a secondary market for a small company's securities, the SEC has permitted the company to post on a bulletin board the names and contact details of persons buying and selling the stock. As a condition, the company may be required to keep records of all quotes, and will itself be prohibited from buying or selling its own shares – for a discussion of the SEC's position on bulletin board trading of Spring Street Brewing, see Coffee (1997: 1215).

The Australian regime for the regulation of online initial public offerings is similar. At its discretion, ASIC may grant conditional exemptions from certain aspects of securities law that would otherwise prohibit the use of electronic prospectuses. Electronic prospectuses lodged with ASIC must be accompanied by a paper version containing the same information. Offerors must take reasonable steps to protect the electronic version from tampering or alteration, and must restrict the use of hyperlinks to minimize the likelihood that they would distract the investor from the principal document (ASC Policy Statement 107 Electronic Prospectuses, ASC Digest PS7/1575; see Simpson 1998: 49). Prospectuses published, through whatever medium, that do not meet with the approval of securities regulators are thus in breach of the law.

Similarly, although Australian law does not specifically address the provision of investment advisory services over the Internet, ASIC has indicated that investment advice provided to Australian investors electronically must comply with Australian licensing requirements. ASIC will not, however, seek to regulate securities offerings and advertisements on the Internet where they are not targeted at persons in Australia, where they contain meaningful jurisdictional disclaimers that indicate in which jurisdictions offers are available, and where they do not involve fraud or misconduct (O'Sullivan and Damian 1999: 55)

Regulatory responses

The first regulatory actions against online investment schemes in the United States were taken in 1994, when the New Jersey Bureau of Securities acted against an Internet variation on the classic pyramid scheme. In some respects, the advent of digital technology facilitates the task of securities regulators. In the past, when many fraudulent investment solicitations were made personally through telephone calls or written correspondence, some matters only came to official attention through the complaint of a victim. This was complicated by the fact that not all victims of fraud were aware that they had been victimized in the first place, and of those who were, not all were inclined to share their embarrassment.

Today, the most efficient means by which a fraudster can gain access to the gullible is the Internet. But such overtures are more visible than personal contact. In many respects these new technologies allow law enforcement authorities greater access to the fraudster.

Surveillance

Surveillance of cyberspace by regulatory authorities began informally in the mid-1990s and has become increasingly institutionalized. Initially, the SEC established a voluntary program called 'Cyberforce' through which employees surfed the Internet, checking out the postings on investor Usenet groups, and looking for apparent manipulation, misleading statements, and illegal offerings. Since then, the SEC has established a sophisticated surveillance program based on new software technologies and improved search engines (Cella and Stark 1997: 835). NASD similarly monitors the Internet in search of fraudulent activity and materially false information about NASDAQ listed companies (Stirland 1996: 34; Martinson 1997: 17).

In Australia, ASIC established an Internet surveillance program to cover message areas such as news groups and bulletin boards. New software tools that assist in automating the surveillance process have already met with some success. The ASX surveillance system, SOMA, detects trading abnormalities which can then serve as the basis for further investigations and referrals to ASIC (see Simpson 1998: 41).

Regulatory authorities also conduct 'Surf Days' during which they patrol cyberspace for suspected fraudulent activity, and call questionable sites to the attention of the responsible jurisdiction. On 28 October 1998, the SEC (1998b) took proceedings against 44 stock promoters caught in a nationwide enforcement sweep to combat Internet fraud. These promoters had failed to inform investors that more than 235 companies had paid them millions of dollars in cash and shares in exchange for touting their stock on the Internet. Not only had they lied about their own independence, some had also lied about the companies they featured, and taken advantage of sharemarket fluctuations to sell their shares for a fast profit.

Enforcement agencies are also able to establish covert e-mail addresses to 'shop' for fraudulent Internet solicitations and to obtain evidence or information for purposes of enforcement proceedings (Hildreth 1999).

The transparency of securities markets facilitates investigation. Unless fraudsters take extraordinary steps to conceal their identities and location, they are likely to leave clear footprints. Those who solicit funds from investors usually provide instructions for the transfer of funds, or some other contact details which can assist in identifying and locating them. Even those who seek unobtrusively to influence share prices will usually leave tracks. Regulatory authorities may seek to enlist the assistance of service providers in identifying them.

Enforcement

Criminal prosecution is a costly and difficult process, no less so when mobilized against securities fraud than against other offences. Notwithstanding the deterrent effect which a successful prosecution and subsequent sentence may produce, the goal of quickly restoring integrity to markets, and protecting naive investors from further predation in the short term, may often work in favour of less draconian interventions. While decisions on the mobilization of law will vary from jurisdiction to jurisdiction, the circumstances of the case and the availability of alternative sanctions will determine the path along which authorities choose to proceed.

Securities fraud does not always attract a criminal sanction. In the case of simple frauds, the foremost objective of securities regulators may be to put the offender out of circulation quickly and inexpensively. As an alternative to long, costly and often uncertain criminal proceedings, authorities may opt for civil remedies. These will serve the basic function of taking a fraud off the market so to speak, and reduce the possibility of further victimization (for examples of a variety of SEC actions against Internet-related irregularities, see Prentice 1998: 21–5).

Most popular of these, at least in the United States, is the injunction or order, which may be issued directly by the regulator. The earliest of these directed at online investment solicitations were issued in New Jersey in 1994 against chain letters and unregistered securities. These may be accompanied by civil monetary penalties, agreement on the part of the miscreant to cease activities (if he has not done so already), and an undertaking not to engage in similar conduct in the future. The Odulo case of the erstwhile eel farmer illustrates this. Without denying or admitting the charges against him, Odulo consented to an injunction from further violations of federal securities laws. A civil monetary penalty was waived because of his financial circumstances.

To illustrate how one jurisdiction dealt with a recent number of cases, the Internet Compliance and Enforcement Unit of the California Department of Corporations had by 1999 issued 39 Desist and Refrain Orders to a total of 159 persons; filed one civil injunctive action and obtained a preliminary injunction against 26 defendants; and referred two cases for criminal prosecution (Hildreth 1999).

Injunctive remedies are also available in Australia. In February 1999 ASIC obtained an injunction restraining the publisher of a website from providing securities reports on the Internet. ASIC suggested that the publisher would be able to apply for a securities adviser's licence, requiring that he document his relevant experience, training and qualifications.

Of course, there are situations in which the fraud is so egregious, and the losses so great, that a criminal prosecution will be deemed appropriate. In some jurisdictions, regulators have severe penalties at their disposal. The Internet hoax of April 1999, which involved a false announcement of a takeover, carried a penalty of five to 30 years' imprisonment and fines of up to US$1 million. As it happened, the perpetrator was sentenced to home detention, probation, and was required to pay US$93 000 in restitution to investors who sold at a loss before the hoax was disclosed, the sentencing judge observing that the accused had not profited from the hoax and had an otherwise honourable life (Associated Press 1999a). In another case, the SEC and Justice Department prosecuted the chief executive of SEXI for price manipulation that generated nearly US$10 million in ill-gotten gains. A lengthy term of imprisonment was imposed (The Business UK 1999).

But irregularities in securities markets are not always this clear-cut. Prentice (1998) has identified some interesting issues that may arise when companies become the unwitting subjects of misrepresentation. When the misrepresentation is disparaging, it is, of course, in the company's interest to rectify the situation as soon as possible and set the record straight. This is the case regardless of whether the source of the misinformation is a disgruntled employee or a short seller. By contrast, when the misrepresentation is unduly positive and originates from a source outside the company, it may be tempting to delay rectification.

The difficulties faced by companies relate to the lengths to which they might be required to go in order to comply with general regulatory requirements. How often should their website be checked or updated to ensure that the information it contains is correct? What is an unacceptable delay in rectifying misinformation? What is the appropriate course of action for a company which discovers that its website is linked to other websites that may themselves contain inaccuracies about the company in question? To safeguard against some of these difficulties, many companies have begun to engage the services of consultants who scan the Internet for corporate references. Such self-regulatory initiatives will be discussed below.

Perhaps somewhat less ambiguous is the matter of what constitutes investment advice. Regulators in most jurisdictions require that investment advisers be required to hold a licence. But anyone can offer investment advice, and citizens of democratic countries are generally free to express their opinions. Normally the requirement applies to those who receive some compensation for their services. But for the time being, it remains the regulator's call. As one popular Australian investment-related website reads, 'this site now has little in the way of investment help. I was given the choice by the ASC [now ASIC]: get an investment adviser's licence or stop giving investment advice. I chose the latter course' (Pauley 1999).

The question has also begun to arise whether software programs that provide investment advice require registration, and whether publishers of such software should be regarded as investment advisers. It would seem that the answer is no, unless the software contains predictive data or assessments of quality. General investment guidelines do not constitute investment advice, but direct or indirect recommendations may do so (Australian Securities Commission 1999).

Then there is the question of the discourse occurring in bulletin boards and chat rooms. When does a rumour-monger become a criminal? In the United States, one is free to express opinions about the wisdom of a particular investment choice, but if one is paid to do so, the law requires that such arrangements be disclosed. If the opinions in question are knowingly false, and proffered with the intention of influencing the price of a stock and benefiting from the change in price, they may attract criminal liability. The difficulty here, of course, will be proving that the offenders knew the information to have been false, and that they intended to defraud. There is also the problem that one can offer online advice to investors anywhere in the world from regulatory havens – jurisdictions that have no licensing requirements.

Investor education

The best line of defence against securities fraud is an informed investing public. Web pages of major securities regulators contain abundant information about the pitfalls that await the unwary. The SEC Office of Investor Education and Assistance publishes an 'Investor Alert' discussing various forms of online investment fraud. The SEC also posts news releases to relevant newsgroups.

Tremendous opportunities also exist for regulatory agencies to communicate with clients and stakeholders. Agencies are increasingly using their websites to provide public information about fraud risks and questionable practices, including the use of online enforcement alerts (Prentice 1998: 11). In the United States, the Commodity Futures Trading Commission, the federal agency for the regulation of futures trading, posts photographs of suspects on its web page and provides a facility to e-mail information about suspects directly to the agency (Prentice 1998: 12; Commodity Futures Trading Commission 2000).

The North American Securities Administrators Association (NASAA) has an established hotline which enables investors to report instances of suspected Internet securities fraud (the address is cyberfraud@nasaa.org). The SEC web page contains a link to the SEC Enforcement Complaint Center (SEC 2000) By 1999, the SEC was receiving 300 complaints a day about online stock-trading activities, mostly 'pump and dump' schemes aimed at persuading investors to buy shares in speculative or shell companies (companies without any real assets). In Australia, ASIC has issued a number of warnings to consumers about the dangers of relying on information on the Internet without first obtaining a current prospectus.

As we have seen, the practice of day trading, which has attracted a number of investors whose optimism is matched only by their naivety, also requires a degree of investor education, at the very least to inform individuals of the risks involved more fully (see FTC 2000; FSA 2000).

International cooperation

The hackneyed adage that cyberspace knows no frontiers reflects new challenges for securities regulators. Fraudulent investment solicitations and other deceptive practices relating to securities transactions may originate on the other side of the world, though determining precisely where a communication originates can be problematic. According to ASIC, an offer or invitation is made in Australia if it is received in Australia. This means that a fraudulent offer originating in Russia accessible from Australia may constitute an offence under Australian law. In the United States, the offer or sale of securities across state borders or into the United States will trigger SEC jurisdiction (Silverman 1997: 448). In practice, regulators will be required to develop policies to cope with jurisdictional issues. In Australia, ASIC's Policy

Statement 141 makes it clear that it will not seek to regulate offers and advertisements emanating from overseas and accessible in Australia if they are not aimed at people in Australia, have little or no impact on Australian investors, contain a meaningful jurisdictional disclaimer, and there is no misconduct.

But what if these criteria are not met? Establishing jurisdiction is one thing; invoking it is another (Silverman 1997: 449). The offender may lie beyond the practical reach of either nation's law enforcement agencies, and international cooperation is by no means automatic (Picciotto 1997). The challenge lies not in making the law, but in enforcing it.

There are a number of factors that will determine the reach of securities law enforcement. First is the law of the nation in which the miscreant is physically located. Not all jurisdictions offer the same degree of investor protection. What is prohibited in one country may be entirely legal in another. Standards will vary from jurisdiction to jurisdiction depending on what is defined as a security and what constitutes an offering. If the host nation's law does not prohibit the conduct in question, it is most unlikely that the authorities in that nation will be inclined to intervene as there is no basis for mutual legal assistance.

Second, even if the host nation and its laws are sympathetic, the cost of bringing proceedings against an accused who may be situated on the other side of the globe is often prohibitive. Even the extraterritorial application of local laws may be of little assistance if the cost of extradition and subsequent prosecution are disproportionate to the severity of the offence.

To illustrate how jurisdiction may thwart the intentions of national regulatory authorities, consider the following statement of a person who chose to base an online financial newsletter, http://www.stockscape.com/, in Canada rather than in the United States:

> One of the reasons we wanted to base it out of Vancouver is because the regulatory environment in the United States is, well – there are a lot of question marks surrounding it. So you don't know what you can say and what you can't say on the Internet. And in fact there are lot more question marks down there than there are in Canada. And so from a pragmatic point of view, we decided it would be more flexible to base it out of a more open environment which is the case in Vancouver. (Newton 1998)

The solution to the jurisdictional predicament posed by the borderless nature of cyberspace which affects many of the issues outlined in this book can only be achieved through a process of global harmonization (Braithwaite and Drahos 2000). Regulatory authorities have begun to weave a web of cooperation articulating common principles pertaining to cross-border communications (see SEC 1998a and FSA 1998). This takes place bilaterally, as well as under the auspices of multinational organizations such as IOSCO (1997). Areas of cooperation will include exchange of best-practice ideas for investor education, consistency of disclosure requirements, market surveillance technologies, and investigative methods. Regulatory authorities can also advise each other quickly when new forms of illegality begin to emerge so that when criminal innovations diffuse, regulators can be prepared to combat them.

Self-regulatory activities

Self-regulation remains an important bulwark against securities fraud. This can occur on the part of the company whose securities may be the subject of fraud, as well as institutions within the securities industry. The importance of companies remaining vigilant against the possibility of misleading information was noted above. This can be the concern of internal committees concerned with compliance generally, as well as external commercial actors who provide Internet monitoring and third-party audits of corporate websites (Prentice 1998). Companies specializing in reputation monitoring and measurement exist for the purpose (Delahaye 2000; eWatch 2000). As we shall see in Chapter 7, the use of website certification and accreditation is becoming one of the principal ways of guarding against online fraud, particularly in relation to consumer activities.

The digital age has provided securities industry associations and stock exchanges with unprecedented market surveillance capability. In the United States, NASD has a Netwatch facility with sophisticated software for scanning and retrieval. The NYSE 'Stock Watch' is an electronic monitoring system designed to detect unusual trading patterns such as anomalies in price and volume, which may be indicative of manipulation and insider trading. The NYSE Enforcement Division may itself take disciplinary action, or may report the matter to the SEC (New York Stock Exchange 2000).

Nor is investor education necessarily the monopoly of regulatory authorities. Indeed, it is an important aspect of industry self-regulatory regimes. In March 1999, NASD proposed that service providers post a set of caveats for prospective investors, encouraging them to be mindful of their financial situation and trading experience. Investors were warned to be wary of advertisements heralding large profits, and were told they should not trade with retirement savings or second mortgages. In addition, many online trading sites also publish information about the risks of online trading. Further warnings are linked to websites of law firms that specialize in representing victims of securities fraud (Endfraud.com 2000).

Although the risks in using the Internet are now readily apparent, the challenge lies in ensuring that those who are most at risk are made aware of such information and, more importantly, act on it. The complexity of new technologies also requires regulators to forge new strategic alliances with the private sector. What Phillipps (1999) calls 'techno-governance' involves liaison between regulators and other institutions such as online brokers and technology developers. These links are intended to develop new systems for the surveillance and detection of misconduct, and for the identification of new applications, with potential for legitimate exploitation as well as misuse.

Conclusion

The advent of digital technology has provided unprecedented access to prospective investors by small entrepreneurs, and has dramatically increased the amount of information available to small investors. Ordinary investors are now able to buy and sell shares without dealing through intermediaries. While this may enhance the efficiency of securities markets, it also provides opportunities for criminal exploitation. Some sense of the brave new world of securities markets and their vulnerability may be gleaned from the following account of the rise and fall of Omnigene Diagnostics.

> The stock was touted on electronic Bulletin Boards; clues about the company's problems first appeared on its site on the World Wide Web; the details were fleshed out using E:mail, which was also employed to alert regulators, and the company's management took to the Net before the SEC stepped in online. (Eaton 1996: D4)

Stock exchanges have adapted to the opportunities and the risks posed by new technologies in the past. The telegraph and the telephone were once revolutionary. The newest risks lie beyond the capacity of any one institution to control. The integrity of securities markets in the digital age will depend on traditional regulatory institutions, industry participants, and the markets themselves. Whether these can compensate for the reduced role of traditional intermediaries and gatekeepers remains to be seen. Meanwhile, it seems appropriate to heed the advice of securities regulators that investors who become involved with small, thinly traded companies which are promoted in cyberspace should be prepared to lose every penny.

CHAPTER 7

Electronic 'Snake Oil': Deceptive and Misleading Online Advertising and Business Practices

This chapter examines the economic risks that face consumers in the digital age. In particular, it considers the extent to which digital communications technologies are being used in misleading and deceptive ways to advertise goods and services, and what remedies exist for those who suffer loss as a result of such conduct. Its connection with the concept of theft derives from the myriad ways in which fraudsters seek to trick consumers into parting with money by acts of deception. In most cases, an act of fraud simply entails theft carried out by means of deception, and the exchange of information – or misinformation – is the vehicle by which such deception takes place.

The responses to consumer fraud are varied. Following our theme of legal pluralism, this chapter will consider what might be described as hard regulatory responses in the form of civil and criminal action as well as soft regulatory responses based on commercial and business solutions. As in other chapters, we shall see that technology itself may provide a number of ways in which online fraud can be prevented.

The discussion will be confined to advertising directed at consumers other than the provision of investment advice and of information relating to corporate share transactions, which have been discussed in Chapter 6. The risks associated with electronic payment systems will also not be examined, as these have been examined in Chapter 2, as has the use of telecommunications systems for telemarketing (see Grabosky and Smith 1998).

The Antecedents of Electronic 'Snake Oil'

Throughout history, consumers have been targeted by deceptive and misleading practices that have ranged in subtlety and complexity from the barely plausible to the refined and sophisticated. Australia in the nineteenth century was a haven for purveyors of 'snake oil' and other worthless goods and services. The details of some of these cases appeared in an inventory of fraudsters operating in Sydney between 1841 and 1844, known as the

Registry of Flash Men (Hall 1993). Included in the Registry, for example, was the case of Emily Syster, a 'begging letter imposter', who wrote to the bishop and other prominent figures in the community assuming various persone, such as an old officer, a lady near confinement, and a ruined tradesman. All were written in the same hand and thus the fraud was eventually detected.

In contrast to charitable contribution frauds, the perpetrators of commercial frauds seek to prey on the gullibility and greed of consumers and investors. It is, indeed, part of the modus operandi of consumer fraudsters to propose an arrangement of doubtful legality in order to enhance its marketability as 'an offer too good to be true', while at the same time providing a safeguard that those defrauded will not report their victimization to the authorities. In circumstances in which the subject matter of the transaction may be socially 'sensitive' – such as the provision of sexual or gaming services – victims of online fraud may be even less likely to report their experiences formally through fear of attracting unwanted publicity.

As with other crimes of acquisition, deceptive advertising and business practices arise from new opportunities in the marketplace. In times of increased business activity, more opportunities thus present themselves. In the United States in the 1920s, for example, opportunities for fraud were rife. Charles Ponzi – whose name has become synonymous with a certain type of fraudulent investment practice (see below) – established the 'Financial Exchange Company' in 1919 which guaranteed a 50 per cent return to investors within 45 days. The company purported to buy international postage coupons in countries in which the exchange rate was low, and resell them in countries with higher rates. Within six months, 20 000 investors had provided nearly US$10 million. Unfortunately, the dividend paid to early investors came from the money invested by new investors. After an exposé in the *Boston Globe* on 2 August 1920, Ponzi was arrested and convicted of fraud (Rosoff, Pontell and Tillman 1998: 5).

Similar opportunities arose in the more recent financial boom of the 1980s, which led to a number of corporate collapses and associated fraudulent activities (Grabosky and Sutton 1989; Sykes 1994). In the years to come, the opportunities will lie in the digital world of electronic commerce and the Internet.

The Sociology of Online Consumerism

The last few decades have seen substantial changes in the way in which consumers go about acquiring goods and services. In the space of 50 years, for example, grocery shops have been transformed from businesses in which the proprietor made up orders for individual shoppers who waited at the counter for their groceries to be selected and wrapped, into supermarkets in which shoppers are presented with up to 7500 different products to choose from themselves. Some hypermarkets in North America now enable customers to scan barcodes on their own purchases as they are selected and pay for the total electronically on leaving the premises.

The latest developments in electronic retailing let shoppers use the Internet to see images of every item available at the shop in question and to place orders and make payments electronically. Some shops even have Internet sites which enable shoppers to indicate the parameters of their proposed purchase and be presented with a list of products deemed suitable by the store. One is able, for example, at Macy's, to see what is on offer for a Valentine's Day gift for one's partner aged 65 in the price range of US$10 to US$50 (Macy's 2000).

At present, the commercial side of the Internet represents an extensive advertising medium in which individuals and businesses across the globe give information to consumers about the nature of their activities and the goods and services they offer. In the future, when secure payment systems have become refined, online commerce may replace conventional shops for many goods and services.

The development of online commerce has created considerable changes in the nature of consumerism. These relate not only to the manner in which transactions are conducted, but also the nature of the products purchased. In the past, consumers were able to see and feel a product before deciding to buy. They would be familiar with the nature of the item purchased and it would be readily apparent if it was suitable for its intended purpose, or if it was defective in some way. Now, however, online consumers buy digital products that they neither see nor feel, and they are often unable to determine their fitness for purpose until the transaction has been concluded. In one recent case, an e-mail account was purchased from an ISP who simply failed to provide the online facilities requested. The breach of contract was not discovered until after payment had been made, and attempts at rescinding the agreement and recovering the US$120 proved fruitless (Tynan 1999).

Many websites now warn consumers explicitly of the dangers of conducting online business in order to protect themselves against legal liability; see, for example, Motorola's (2000) disclaimer statement of more than 1700 words, which concludes, 'If you do not agree to these terms, do not use the Site'. Bargaining power, has, accordingly, been shifted considerably in favour of the online merchant, and the principle of *caveat emptor* reaffirmed.

Electronic commerce has also reduced many of the psychological elements involved in the bargaining process. Meeting a retailer online and negotiating an agreement now gives the parties no opportunity to assess each other's nature and reliability because the normal social and business cues are absent. These include general appearance, facial expressions, verbal or body language, voice, dress, demeanour or apparent authority. While these cues may not be essential to the meaning and intent of a transaction, they do provide useful contextual information for one or other party (Research Group into the Law Enforcement Implications of Electronic Commerce

1999: 62). In the future, when greater bandwidth may allow interactive visual online trading, some of these social cues may be reinstated, enabling merchants and their customers at least to see and hear the type of person they are dealing with.

Online shopping also allows consumers to deal directly with manufacturers of products, thus bypassing retailers and distributors. Although this makes transactions simpler, some manufacturers might not have systems in place to deal adequately with consumers such as those commonly employed by retailers. Airlines, for example, now permit customers to buy tickets online without the involvement of travel agents (although some online travel agents also exist). Although this service is prompt, customers might not receive the additional services that travel agents usually offer, such as advice on insurance, the special conditions that attach to some tickets or the provision of travel itineraries.

In order to enhance the operation of online commerce, it has been predicted that new forms of 'brokers' will emerge in the digital marketplace to conduct such activities as customer acquisition, shipping and delivery, credit checking, handling of returns, processing payments, and support and service (Cope 1996: 112). These aspects of retailing were previously dealt with by those in the production and supply chain, but in the online environment where consumers deal directly with manufacturers, these functions may be overlooked. Online commercial brokers of this kind will undoubtedly become an important feature of conducting business electronically in the years to come.

An additional aspect of online commerce concerns the global nature of transactions. Products such as software may be manufactured in one country, advertised from another, and purchased in a third. A consumer in Australia may read an advertisement on the Internet that has been created in the United States but transmitted through various other countries; the product may then be dispatched from England, with payment eventually made to a Japanese manufacturer. Not only does this make the transaction more complicated in the sense of entailing multiple agreements, but it also creates considerable problems if the product is defective and the consumer wishes to seek redress. This problem is, of course, not new, though electronic commerce contemplates the possibility of such international transactions taking place regularly. Because consumers are unable to deal directly with a vendor located geographically nearby, far greater trust is engendered in the transaction between parties who may be in different countries and never able to meet. Where that trust has been breached, the possibility of redress is inevitably remote (see Rothchild 1999: 896).

Computer technologies are also making the process of advertising and marketing much more extensive and efficient. Traditional technologies of marketing such as door-to-door visiting, mail, the telephone, and facsimile machines are now being replaced by e-mail and the Internet. The direct marketing industry, for example, makes extensive use of databases containing detailed information about prospective consumers which can be used most effectively to target advertising. This has given rise to so-called 'database marketing' and a trade in databases that contain personal details about consumers and their purchasing preferences. One company, Acxiom, based in Little Rock, Arkansas, but with international operations, is one of the largest data warehousing companies in the world, recording sales of A$750 million in 1998 (Barker and Johnson 1999). Having identified potential purchasers, database marketers are now able to employ 'push technologies' in which information is automatically delivered to consumers, rather than having to wait for consumers to request or 'pull' the information they want (Highland 1997). The privacy implications of such activities are, clearly, profound and are discussed in Chapter 10.

The Nature and Extent of Online Fraud

Traditional misleading and deceptive practices generally fall into one of four categories: pretending to sell something you don't have while taking the money in advance; supplying goods or services which are of a lower quality than the goods or services paid for, or failing to supply the goods and services sought; persuading customers to buy something they don't really want through oppressive marketing techniques; and disguising your identity in order to perpetrate a fraud. These practices, like those relating to the marketing of the colloquial 'snake oil', have always existed and a fool today is no less easily parted from his money than he was in the pre-digital age.

In the traditional marketplace, many of these risks could be reduced by consumers being able to inspect the product they wanted to buy before parting with their money. Merchants were also able to receive cash in hand before parting with their goods. In the online world, the risks to both merchants and consumers have increased greatly as goods and services generally cannot be inspected before a transaction is concluded. Cash payments have also been replaced by electronic transfers of funds, with the potential for transactions to be concluded instantaneously. While increasing the speed and efficiency of business transactions, these developments also substantially increase the opportunities for fraud and illegality.

Deceptive practices are easier to carry out electronically than using traditional practices because it is a fairly simple task to make oneself appear legitimate. Through the use of sophisticated web page designs, it is much easier to disguise intent and to present a scheme in a positive light, and the pool of potential victims is greatly enhanced by the use of extensive e-mailing techniques. In addition, the financial outlays required for online merchants to set up business are much less than with traditional forms of marketing, as offices are not required and stationery and staff costs are reduced. In the electronic marketplace it is also easy for fraudulent businesses to disguise their true identities, to hide, to shut down or to move to a new location when the illegal nature of their activities starts to become apparent (Wahlert 1998: 8).

The Internet also creates a psychological environment conducive to fraud. The attractiveness of web pages and the constant pressure to carry out activities immediately – exacerbated by timed Internet access contracts with ISPs – means that online shoppers and investors may feel compelled to make decisions quickly and without an opportunity for reflection. The consequences of error may also be immediate. Once payment has been authorized electronically, it is often difficult to rescind the arrangement, and often practically impossible to recover one's funds. Required contractual cooling-off periods thus sometimes become meaningless in the pressurized online environment. There is also a reluctance on the part of many to admit the limitations on their technological understanding and expertise, allowing them to enter into agreements whose detail and implications they may not fully understand.

The Internet also enables the traditional boundaries between information, advertising and entertainment to blur. Consumers can be tricked into believing that the facts they have been supplied with are objective and reliable, when in fact they are merely advertising puffs. The use of such techniques, accompanied by exotic graphic and audio displays, has proved to be an effective marketing strategy when directed at children. The use of hyperlinks embedded in information pages or search engine sites which direct users to advertising material can also be an insidious promotional technique in which readers are asked to believe that the linked material is of an objective, factual nature when in fact it is merely supportive of some specific advertised product. Hyperlinks have also been used to direct consumers to other sites which are covertly operated by the same business as the original site, a practice currently being investigated by the FTC in the United States (Clausing 1999a).

The extent to which deceptive and misleading practices take place online is difficult to quantify precisely, though some estimates have been attempted. In the United States, over 18 600 complaints were registered on the FTC's fraud database 'Consumer Sentinel' in 1999, more than double the number in 1998 (US Department of Justice 2000). One commentator has estimated that as much as 10 per cent of online commerce may involve consumer fraud (Rothchild 1999: 897, n. 11). In 1999, a survey of Internet trading was conducted at the University of Utah, coordinated by Consumers International and funded by the European Union. Representatives of those groups bought more than 150 items from websites based in seventeen countries, and then tried to return them. It was found that 8 per cent of the items ordered never arrived; many websites did not give clear information about delivery charges; a minority disclosed whether the laws of the seller's country or the buyer's country would apply in the event of a dispute, and only 53 per cent had a return policy. In addition, only about 13 per cent of the sites promised not to sell customers' personal data to a third party and only 32 per cent provided information on how to complain if there was a problem with a transaction (Clausing 1999b). In Australia, ASIC and the

ACCC have both found an increase in the extent to which the Internet is being used for deceptive practices in recent years.

What, then, is the nature of these misleading and deceptive practices, and how have they been adapted for use in the digital age?

Types of Misleading and Deceptive Online Practices

Advance fee schemes

The gist of these schemes is to trick prospective victims into parting with funds by persuading them that they will receive a substantial benefit in return for providing some modest payment in advance. The characteristics of this type of fraudulent scheme usually entail enlisting the services of the prospective victim to assist in an activity of questionable legality, thus providing some assurance that the victim, once defrauded, would be unlikely to report the matter to the police. The victim would, rightly, be apprehensive that she had aided and abetted some criminal activity and would also be reluctant to make public the fact of her gullibility, particularly if adverse media coverage were a possibility. Thus the offender is able to carry out the scheme repeatedly, sometimes in respect of the same victim, while police are faced with difficulties in finding witnesses and securing evidence.

Pyramid schemes

Pyramid sales schemes have the primary purpose of enlisting individuals to earn money by recruiting other people. They operate through the use of chain letters, mailing lists, money-making clubs, multilevel compensation plans for low-value publications, and cooperative marketing networks that make use of overvalued or useless phonecards or credit cards. Victims are induced to subscribe funds in the expectation that they will receive payments for introducing further participants directly or through others. Saturation point is generally very quickly reached and later recruits have little chance of recovering their money.

Pyramid schemes are to be distinguished from direct selling schemes, including multilevel marketing schemes, which recruit participants for the purpose of selling products but require any rewards to come directly from product sales. These are a legitimate form of business activity in most countries.

In Australia, section 61 of the *Trade Practices Act* 1974 (Cwlth) makes it illegal to promote or participate in a pyramid selling scheme. The Act prohibits marketing schemes that give rewards for the introduction of new participants (s. 61), misleading and deceptive conduct (s. 52), misrepresentations about business opportunities (s. 59) and misrepresentations about employment opportunities (s. 53B). Penalties include fines of up to A$200 000 for a company, or A$40 000 for individuals, as well as other legal remedies of injunctions, damages, corrective advertising and legal costs.

In the United States, the FTC and various state authorities have taken action against the creators of pyramid marketing schemes conducted online (see Rothchild 1999: 904). One of the earliest and largest to be uncovered involved three individuals in Washington who conducted a business that offered a minimum income for investors of US$5250 a month in return for an advance payment of between US$250 and $1750. The FTC found that between US$6 million and $11 million had been obtained from between 30 000 and 40 000 consumers in 60 countries. The case was settled on the basis that the company repay up to US$5 million. In addition, the defendant's website had details of the court order displayed and a hyperlink created to the FTC website (*Federal Trade Commission* v *Fortuna Alliance* W.D. Wash. Filed 23 May 1996; Da Silva 1996; Starek and Rozell 1997: 689–90).

In March 1999, the FTC commenced 33 actions against 67 defendants promoting Internet pyramid schemes. It also launched a 'sweep' of the World Wide Web to locate sites that might be hosting illegal pyramid marketing scams. In one pyramid operation cited by the FTC, Five Star Auto Club Inc. of Poughquag, New York, promised online consumers an opportunity to lease their 'dream vehicle' for free while earning between US$80 000 and US$180 000. All they had to do was pay an annual fee and $100 in monthly payments and recruit others to join. It was alleged, however, that those who signed up received no free lease on a car and no annual earnings. The FTC filed a lawsuit in the US District Court in White Plains, New York, against the Five Star Auto Club, seeking a permanent injunction and consumer redress, with the federal court ordering the business to close pending resolution of the case (Associated Press 1999a; see also Hogan 1987).

In Australia, the ACCC has also investigated a number of Internet-based pyramid schemes. One involved the Canadian Global Interactive Investments Club, which advertised the fact that offshore banks would issue low-interest Visa cards via the Internet with the promise of no credit checks or income verification. The scheme was actually an international pyramid scheme, in which applicants were encouraged to recruit others in return for A$25 each. Another scheme investigated was the 'Golden Sphere' pyramid selling scheme set up by a Vanuatu company and operated from Queensland. The ACCC sought to restrain promoters of the business and instituted a class action on behalf of Australian participants who had lost money in the scheme (ACCC 1997b).

Another scheme investigated involved the Destiny Telecomm Telecard Scheme, which operated from California. This scheme involved consumers purchasing a phonecard for US$100 for international, STD and/or mobile telephone use. If those consumers introduced further customers to the scheme, they received a commission, the amount of which increased according to the number of customers introduced. The ACCC was concerned that consumers were providing credit card details and debit authorities to an overseas company whose activities may have breached the referral selling and pyramid selling provisions of the *Trade Practices Act*. The ACCC has also recently been liaising with Italian authorities in its investigations of the 'Pentagono' pyramid selling scheme. It appears, therefore, that the Internet has provided an effective medium in which pyramid schemes are able to flourish.

Ponzi schemes

As we have seen, Charles Ponzi, the post-World War I financier, devised a fraudulent means of paying those who first provided investment funds by using the funds of later investors. The Internet is now being used to advertise such operations. As we saw in Chapter 6, franchises for ATMs offering attractive returns have in fact simply been Ponzi schemes in which moneys paid to the early investors are taken from later investors rather than from profits, thus preventing later investors from reaping any rewards from their investments (Hogan 1997: 174–5).

Chain letters and bulk e-mail

Chain letters require the recipient to send a small amount of money to each of four or five names on a list, replace one of the names with the recipient's name, and forward the revised list to others. It is claimed that those participating may earn up to $50 000 in three months. Apart from the fact that this sort of deal infringes local laws, those involved invariably receive nothing in return for their advance payments.

The use of e-mail has enabled such schemes to flourish as databases of names and addresses can be compiled and messages sent in bulk. Other schemes that make use of e-mail set out to persuade users to buy databases containing e-mail addresses in order to target potential customers for their products. Not only are the lists provided of poor quality, thus resulting in few products being sold, but most ISPs prohibit the use of bulk e-mailing. In some jurisdictions, the practice is also illegal (FTC 1999).

One particularly endemic form of advance fee fraud which emerged during the 1980s involved a group of expatriate Nigerians who are believed to operate from cells in the United States, Britain, Canada, Hong Kong, Japan and other African countries with the assistance of confederates within West Africa (Gup 1995: 120–5; Smith, Holmes and Kaufmann 1999). The frauds that have been discovered to date have taken a variety of forms. All have entailed victims being approached by letter, or recently, e-mail, without prior contact. Victims' addresses are obtained from telephone and e-mail directories, business journals, magazines or newspapers, and letters are invariably handwritten, often with counterfeit postage stamps being used, resulting in their seizure by postal authorities. They generally describe the need to move funds out of Nigeria and seek the assistance of the victim in providing bank account details in an overseas country and administration fees needed to facilitate the transaction. The victim is offered a commission, which could be up to 40 per cent of the capital involved. Capital sums of between US$20 million and $40 million are often mentioned, thus creating a potential reward for the victim of up to US$16 million. An advance payment is usually required that could total US$50 000, which represents the amount stolen.

The US Secret Service estimates that since 1989, US$5 billion has been stolen in this way from victims throughout the world. Between August and November 1998, Australia Post, in Sydney alone, confiscated 4.5 tonnes of advance fee correspondence which had counterfeit postage, amounting approximately to 1.8 million items (Smith, Holmes and Kaufmann 1999). One advance fee e-mail message recently received in Australia came from Benin from a citizen of Zaire who claimed to be in possession of two suitcases containing US$10 million which he took with him when President Mobutu fell (*Sydney Morning Herald* 14 July 1998). E-mail has proved to be an effective way of disseminating advance fee letters as the true identity of the sender is easy to disguise and original supporting documentation cannot be checked for authenticity.

Business opportunity schemes

Business opportunity schemes offer consumers the chance to make a significant amount of money in a small amount of time, often by working from home. Examples include schemes in which consumers are given information about how much money they will make and asked to buy a booklet or information page that will tell them how to do it. In many cases, the information given simply tells victims to use the same scheme on other consumers. Another variant involves so-called 'envelope stuffing' in which consumers are paid to print up letters or reports and send them to large numbers of other consumers. The most common concerns about these ventures are that they often dramatically overstate the achievable earnings, and are vehicles to sell large numbers of low-quality or worthless products to consumers who are then unable to onsell them (see Rothchild 1999: 905).

Because many Internet users operate from home, the Internet offers great opportunities for those conducting fraudulent home-based work schemes. Those targeted are often people who have been made redundant or who have been injured and have compensation payments available to invest in a new venture. One example concerned a 42-year-old man in Chattanooga in the United States who had lost his job as an aircraft welder and who also had degenerative arthritis of the spine. He found a home-based work opportunity on the Internet advertised by an organization which claimed it had thousands of people earning US$4000 a month using home computers. He invested more than US$16 000 in the venture and was given an outdated computer system which he attempted to use to operate the company's automated time or weather service and telephone reminder business. The FTC took proceedings against the organization and recovered US$5 million, which was distributed to the 15 000 victims of the operation. In 1995, the US Council of Better Business Bureaus received more than 215 000 inquiries about work-from-home schemes (Hogan 1997).

In Australia, section 59 of the *Trade Practices Act* creates offences of making false, misleading or deceptive claims about business opportunities, including predictions of future earnings, and has been particularly useful in helping to eradicate certain franchising schemes in which the expected financial returns are greatly exaggerated (see *Ducret* v *Colourshot Pty Ltd* (1981) 35 ALR 503).

Auctions

Auction fraud in 1998 accounted for two of every three complaints made to Internet Fraud Watch, an online fraud-reporting system created by the National Consumers League in the United States. Some of the main problems with online auction sales lie in inadequate user authentication procedures for those who participate, and non-delivery of the goods purchased (Rothchild 1999: 907, n. 48). Other problems concern the auctioning of illegal items, which have even included human organs. In the United States, eBay now offers free insurance for those who participate in its online auctions in order to enhance the security of transactions. It also acts promptly to deal with illegal auctions and withdrew the sale of a human kidney for US$5.7 million after this was brought to its attention. In Australia and New Zealand, Yahoo commenced an online auction business in which revenue comes from advertising rather than from users, thus offering some further protection (Heywood 1999).

Prizes and lotteries

A number of misleading and deceptive practices have also arisen in relation to competitions and lotteries, activities that are closely regulated by most countries' local laws. Some are simple advance fee schemes in which consumers are given the opportunity of gaining access to a winning ticket if they send an advance payment, commonly described as an administration fee. Others attempt to pass themselves off as legitimate licensed lotteries when in fact those who buy tickets are unable to win anything. Some require participants to travel overseas at their own expense or purchase other substantial goods or services before they can collect their tickets.

In many countries such practices are illegal whether or not they emanate from the Internet. In Australia, for example, section 54 of the Trade Practices Act makes it illegal to make claims about prizes of lotteries which are false, misleading and/or deceptive; to offer gifts or prizes with the intention of not providing them, or of not providing them as offered; or to provide a price for entry in a lottery which is not the full cash price.

Non-delivery and defective products and services

Misleading advertising on the Internet may also involve the merchant failing to deliver the goods when and where requested, or at all. This problem is exacerbated in global commercial transactions where long-distance delivery of goods is involved, sometimes entailing customs clearance and the payment of importation taxes. Alternatively, defective goods may be delivered or the wrong ones supplied, in which case consumers need to arrange for return and the refund of the price paid and any expenses involved. Many of these problems arise from the fact that the goods cannot be physically examined online in the first place. Problems are also created by the immediacy of some transactions in which goods or services are obtained at exactly the same time as payment is authorized (e.g. the purchase of software downloaded after providing a credit card number). In such cases, the traditional methods of stopping payment on a cheque or withdrawing authorization for a credit card purchase are not available.

Provision of Internet services

As consumers continue to increase their use of the Internet, so the number of complaints about ISPs has increased. In Australia, complaints to the ACCC have included allegations of overbilling, inadequate detail when billing, failure to supply technical support and other services as represented, failure to connect consumers to the Internet as agreed, not honouring requests to disconnect, the need to have a credit card to obtain services, attempts to avoid consumers' legal rights, and misrepresentations about the speed of Internet access and the experience of the ISP (ACCC 1999). In the United States a particular problem being investigated by the FTC has involved allegations of failure to disclose the terms of free trial offers when offering online services (Starek and Rozell 1997: 692).

Cable decryption kits

The introduction of cable television has created enormous opportunities for theft of services through the use of devices that decode encrypted transmissions. As early as 1993, for example, the cable television industry in the United States lost nearly A$6.6 billion from theft of premium and basic services (Peachey and Blau 1995: 2–5).

The Internet is now being used to advertise cable television decryption kits that enable consumers to obtain cable television without paying contract fees. Not only do such kits seldom work but they also require consumers to break the law by stealing cable services. Already a number of prosecutions have taken place of those who manufacture or sell decryption kits, and in Britain, following the decision of the House of Lords in *BBC Enterprises* v *Hi-Tech Xtravision* ([1993] 3 All ER 257), the *Broadcasting Act* 1990 was amended to create an offence of making, importing, selling or letting for hire any unauthorized decoder for encrypted television broadcasts (s. 179). In the future consideration may need to be given to creating a new offence of advertising such equipment for sale.

Computer products and services

In the United States prosecutions have been taken by the FTC in respect of Internet businesses that offer computer memory chips for sale but fail to deliver the goods once ordered by customers within a reasonable time. In one case the FTC was able to obtain court orders that prohibited the misrepresentations and required the payment of US$5000 in compensation (*Federal Trade Commission* v *Brandzel* N.D. Ill. Filed 13 March 1996).

In the digital age, it has also become possible to purchase a variety of digitized products by downloading them electronically. The Internet provides a comprehensive advertising medium for these products, often with online samples being provided free of charge. Once payment has been transmitted to the merchant, the purchaser is able to download the complete product along with appropriate licences for its use. If, however, the software has been illegally copied, or is defective in some way, the consumer is often disappointed, having paid for a product that cannot legally be used.

The regulation of misleading practices relating to computer games and audiovisual material generally comes within the scope of the laws that govern the advertising and dissemination of publications generally. Illegality may also occur where unclassified materials are sold. In Australia, for example, specified publications, computer games, films and online services are all governed by Commonwealth and state and territory legislation. The *Classification (Publications, Films and Computer Games) Act* 1995 (Cwlth), for example, which has counterparts in other Australian jurisdictions, regulates the content of classified material of this nature. If unclassified material is sold online, then the act of sale may itself be illegal as well as the product sold being unable to be used legally.

Sexual services

An area of online commerce which has developed prolifically is that which concerns the provision of pornographic images and sexual services. The present discussion is not concerned with questions of if and how such content should be regulated, but rather the presence of deceptive and misleading practices in the dissemination of such material. Owing to the sensitive nature of the content, victims are reluctant to report such schemes.

One example was of a company that advertised 'free' erotic photographs on the Internet. In order to see the images, the user was required to download software which, once installed, took control of the user's modem, cut off the local ISP, and dialled a number in the former Soviet Republic of Moldova in Eastern Europe. The line remained open until the computer was turned off, resulting in the user incurring large international telephone charges which were shared between the fraudster and the Moldovian telecommunications company. The fraud was detected through regular surveillance of customers' telephone accounts and the FTC was able to obtain an order requiring the defendants to place US$1 million in an escrow account pending resolution of the case (*Federal Trade Commission* v *Audiotex Connection Inc.* E.D.N.Y. Filed 13 February 1997; Anonymous 1997; Grabosky and Smith 1998: 139; Starek and Rozell 1997: 692–3; Rothchild 1999: 909, n. 63).

Misleading credit and loan facilities

In various jurisdictions, misleading advertising by finance-brokers or money-lenders is specifically proscribed. Some legislation also requires moneylenders to disclose their licence number and give details of the interest rate of certain transactions (see Goldring et al. 1998: 307–53).

In addition to acts of unlicensed practice, some misleading practices have involved loans being offered for homeowners with inadequate equity, or credit being offered for those who are unable to establish an adequate credit history. In fact all that is provided is the name of a lending institution, often one that charges above-average rates of interest.

Consumers with poor credit records may also be told that for a fee, paid in advance, their bad credit report can be cleared or a completely clean credit record given to them. In the first Internet case prosecuted by the FTC in the United States in 1994, the defendant had advertised a credit repair kit on the Internet that would enable users to create an entirely new credit history. The Commission obtained orders prohibiting the defendant from making further similar representations and requiring him to pay compensation (*Federal Trade Commission* v *Corzine* E. D. Cal. 21 November 1994). Similar online credit repair schemes have since been prosecuted by the Commission (Starek and Rozell 1997: 687; Rothchild 1999: 906, n. 44).

Health and medical products

'Scientific breakthroughs', 'miraculous cures' and 'secret formulas' have always been used to sell various health and medical products, often with no validity and sometimes with counterproductive effects on health. In one recent case, for example, an American television retailer was ordered to pay a US$1.1 million civil penalty for broadcasting advertisements for a variety of skin care, weight loss and premenstrual syndrome/menopause products which contained unsubstantiated claims of effectiveness (*Federal Trade Commission* v *Home Shopping Network* [1999] *Corporate Crime Reporter* 19 April: 5–6).

The Internet has proved to be an effective means of advertising health and medical products and as legislative controls tend to be local, there are often problems associated with taking action against those responsible. In the United States the National Council against Health Fraud investigates false and misleading claims relating to health products on the Internet and advises local regulatory agencies to take action where appropriate. One company was prosecuted in Illinois for advertising Germanium Sesquioxide as a drug that was said to be able to lower cholesterol, reduce arthritic pain, treat AIDS and cure cancer. The drug had, however, been banned by the Food and Drug Administration after it was found to cause kidney damage, coma and death (Hogan 1997; Rothchild 1999: 906, n. 47).

In another case, a car dealership company based in Las Vegas placed notices on the Internet indicating that a subsidiary company had developed a cure for HIV. The notice, which was repeated, led to a rapid increase in the value of the company's shares on the stock market. On 22 July 1999, however, the SEC suspended trading in the company's shares pending the hearing of fraud allegations (Romei 1999).

In most jurisdictions specific legislative controls have been enacted to proscribe misleading and deceptive advertising of medicinal products and health cures. Cures for cancer, for example, are generally prohibited by local legislation, which has not, unfortunately, prevented them from being widely disseminated on the Internet (Varney 1996). Registered health care providers, such as medical practitioners, are also subject to specific controls which ensure that they do not engage in misleading or deceptive advertising, or advertise products in which they have some commercial interest (e.g. in Victoria, Australia, see *Medical Practice Act 1994* (Vic), s. 64(1)). In the age of telemedicine, these types of conduct are already present and may be expected to increase substantially.

Educational qualifications

The Internet is now being used for higher-education purposes, with entire courses being offered online. Students simply enrol in the course and receive educational materials electronically and communicate with lecturers via e-mail. Seminars may be conducted through the use of chat rooms and newsgroups, and assignments can be sent electronically with examiners' comments provided by return e-mail using 'revisions' software. There have, however, been instances of online universities failing to provide recognized, or indeed any, valid qualifications, or occasionally failing to deliver any educational programs at all.

In the United States, nineteen people were defrauded when a website calling itself 'Loyola State University' advertised Bachelor's, Master's, and doctoral degrees for between US$1995 and $2795. All that candidates were required to do was to send details of their life experiences along with the payment and they would receive their degree within a month (Denning 1999: 132; Rothchild 1999: 908, n. 61).

Unsolicited and unwanted goods and services

Unsolicited advertising

Traditionally, there were few controls on advertising conducted by mail, and direct marketers inflicted a barrage of advertising material on unsuspecting and often unwilling recipients. The electronic equivalent, known as 'spamming' (after a comedy sketch on *Monty Python's Flying Circus*), entails the same idea carried out through the use of e-mail (Clarke 1998). Its future equivalents may be even more invasive, with self-opening attachments that could carry viruses into the recipient's computer hard drive, causing damage and loss. In one example of this, a Portuguese hacker developed a technique by which users who conducted searches for legitimate websites – such as Internet game sites – were presented instead with pornographic images from

Australian sites and were unable to remove these without turning off their computers. The user was, accordingly, compelled to receive salacious advertising material from sites such as www.tabooanimals.com (Labaton 1999). Sending unsolicited online communications may cause serious disruption to service providers and users, in addition to the considerable cost incurred in online time taken to read or deal with messages. Receiving a large file containing complex graphic visual or audio material can take time, even if the attachment is not opened.

In some jurisdictions in the United States it is illegal to send unsolicited facsimile and e-mail messages unless certain conditions are satisfied. In Nevada, for example, legislation makes a person who sends an e-mail containing an advertisement liable to the recipient for damages (US$10 per e-mail, plus attorney's fees and costs), unless there is a prior business relationship or consent, or the e-mail states it is an advertisement and provides the name, street address and e-mail address of the sender, and a notice instructing the recipient how to decline the receipt of further e-mail advertisements (Rothchild 1999: 935, n. 161). In the future, such legislation may become common internationally, although enforcement across borders will remain a problem.

Bait advertising

Bait advertising is the offer of a product or service for sale at an enticingly low price in order to sell some other more expensive product or service, or advertising a bargain that does not exist, in order to attract customers to do business with the merchant. In Australia, such advertising is prohibited by section 56 of the *Trade Practices Act* 1974 (Cwlth). In one recent case, an American sought to purchase a compact disk from an English company which advertised it for sale on the Internet at £8.99. After requesting the item, but before the order was accepted by the seller, the price was increased to £12.99. Complex legal problems arose for the purchaser, who tried to hold the seller to the original lower price. Apart from legal costs being prohibitive, the relevant misleading advertising laws applied only to consumers who read the advertisement in England, not in the United States (Kaplan 1999).

Inertia selling

Another objectionable practice is inertia selling or sending unordered goods to consumers and billing them in the hope that they will accept the goods and pay the bill without question. Various statutes now make such practices illegal, and they would probably apply where electronic goods or services are provided to online consumers without their request. One could imagine software being provided as a self-opening attachment to an e-mail message which would then be billed. Similarly, requiring payment for access to Internet sites could amount to a form of inertia selling of the service in question.

Viruses

Another form of advertising which involves bulk mailing of unsolicited material has developed recently with the advent of electronic Macro viruses. The Melissa virus, which was transmitted in an attachment to an e-mail message, was thought to have been generated by an individual seeking to disseminate advertising messages on a wide scale. When a user opened the file, labelled list.doc, attached to the e-mail, it executed a Microsoft Word macro which searched the recipient's hard drive for an address list produced by one of two Microsoft e-mail programs, Outlook or Outlook Express. It then copied the message to the first 50 names in the address book, giving the appearance of having been sent by someone known to the recipient. It was thought that the program was written in an attempt to promote pornographic sites by automatically spreading them across the Internet in a way that made each message appear to have come from an acquaintance of its recipient (Richtel 1999a).

Identification deception

The advent of online commerce has also created new forms of illegality which are less likely to occur in traditional marketplaces. Many consumers, for example, now have great difficulty in identifying those with whom they do business. Some merchants may intentionally disguise their identity through the use of remailing facilities in order later to defraud customers and avoid detection (see Rothchild 1999: 927). Others may simply neglect to provide accurate and verifiable information.

The technology of the Internet makes it fairly simple for users to disguise their identities. E-mail and Internet addresses may be manipulated by including misleading details, or the source of a message may be made anonymous or changed so that it appears to be coming from another user. Similarly, there is no way of knowing the commercial affiliations of those on the Internet. Referees for businesses or products might in fact be individuals employed specifically to indicate their approval of the venture or product in question.

Businesses might also choose legitimate-sounding names in order to improve their credibility or include domain names which are misleading. There has recently developed a practice in the United States and Canada of some businesses adopting domain names containing the names of Australian cities in order to improve their marketability and credibility, despite the fact that they have no connection at all with Australia.

In one case investigated by the ACCC (1997b), an Internet trader used the same domain name as another trader (the original bearer of the name), but with a <.com> suffix, as opposed to the <.net> suffix of the original site. The confusion created as to the identity of the actual proprietor of the site allowed consumers to be misled or deceived. The <.com> site did, however, include an inconspicuous notice stating that the site should not be confused with the <.net> site of the same name, though this could easily have been overlooked by those visiting the site. In another case, the ACCC has instituted proceedings for declarations and injunctions in the Federal Court of Australia against a company for using a domain name almost identical to an official domain name registering authority based in the United States (ACCC 1998: 67).

Concerns over the misleading use of domain names are being addressed by the introduction of systems of registration such as that administered by .au Domain Administration (2000). This company's membership rules specify that companies must be identified by an Australian Company Name and that non-incorporated entities must be identified by Australian Registered Business Name. Individuals or organizations without contact details that include an Australian paper-mail address and an Australian telephone number are not eligible to receive a domain name. Post office box numbers are also not acceptable as specified addresses.

Hard Regulation

The regulation of advertising and marketing is a relatively new phenomenon that was gradually introduced as the twentieth century progressed. One early attempt to address the question of unethical practices in advertising took place in Sydney between 29 August and 3 September 1920 when the Second Convention of Advertising Men in Australasia was held. The convention aimed 'to raise the status of advertising, to declare its principles, and to improve its practice' (Waller 1992a). It led directly to the establishment of the Advertising Association of Australia and New Zealand, the first national body representing the advertising industry in Australia (Waller 1992b).

Consumer advocacy groups, which emerged in the 1970s, tended to demand strict legal prohibition of unethical practices (the 'hard regulation' approach), while those within the business community felt that self-regulation through the use of codes of practice was just as effective ('soft regulation'). With the introduction of new communications media – the telephone, radio, television and later the Internet – debate has continued on the appropriate form regulation should take.

Rules for determining acceptable practice relating to the sale of goods have taken hundreds of years to develop in the common-law world. In England the so-called 'law merchant', which was a body of rules distilled from the customs adopted by those in the marketplace, was codified in the *Sale of Goods Act* 1893. The rules embodied in this legislation formed the basis of current practice in those legal systems derived from English common law, though the detailed provisions that now apply have been considerably modified and supplemented with new provisions.

Most of the advertising content that appears on the Internet is, to use the legal term, in the nature of an invitation to treat, or mere puffery. Only if interested consumers respond by disclosing their personal details, which may include a name, address and credit card account numbers, will a formal offer to purchase be transmitted which, if accepted and supported by consideration, will give rise to a legally binding agreement (*Carlill* v *Carbolic Smoke Ball Co* [1892] 2 QB 484; see Davies 1997).

Those who display misleading or deceptive advertisements on the Internet will generally only be held liable if the objectionable content forms part of the terms of the agreement. This may give rise to a right to rescind the contract or sue for damages. In this sense, the use of the Internet raises legal issues which are substantially the same as those that arise out of paper-based advertisements and contracts. There are, however, particular evidentiary and forensic difficulties associated with establishing what transpired between the parties to an electronic transaction.

The first and perhaps most problematic legal issue concerning electronic commerce is the attribution of electronic communications, or being able to prove who the contracting parties were and what was communicated between them. Systems of electronic commerce need to ensure that fraudsters are unable to deny having sent illegal communications which in fact they did, and also that innocent parties are able to prove that illegal communications were not sent by them when in fact they were not.

These problems have been addressed in a wide range of legislative models adopted internationally in recent years to ensure that contracts entered into electronically are binding and able to be proved in subsequent legal proceedings (see McCullagh 1998; Sneddon 1998). In Australia, for example, the federal *Electronic Transactions Act* 1999 (Cwlth) provides that the legal requirements for information to be in writing, signed, produced, recorded, and retained can all be satisfied through purely electronic means. Provision is also made for determining the time and place of the dispatch and receipt of an electronic communication. Finally, the Act provides that the purported originator of an electronic communication will be bound by it only if the communication were sent by the purported originator or with the authority of the purported originator. Laws such as these aim to facilitate electronic commerce by providing a regulatory framework in which other laws operate.

Other legislative solutions are also being devised to deal with the legal and evidentiary problems associated with proving an intention to defraud in cyberspace. Originally problems existed in establishing deception where it was only a computer that was deceived and not a human actor. In Australia, Chapter 3 of the Model Criminal Code contains a number of provisions that deal with such problems. The definition of 'document' in section 19.1(1) relating to forgery and offences involving false documents, for example, includes 'a disc, tape or other article from which sounds, images or messages are capable of being reproduced', while section 19.1(2) provides that 'a reference to inducing a person to accept a false document as genuine includes a reference to causing a machine to respond to the document as if it were a genuine document'. In addition to laws regarding actionable misrepresentation of the terms of a contract, most countries also have local consumer protection legislation which governs a wide range of misleading and deceptive practices. In Australia, one of the first consumer protection statutes to be enacted was the *Book Purchasers Protection Act* 1899 (NSW), which sought to regulate the conduct of itinerant merchants who engaged in door-to-door sales. Since then, more restrictive legislative regimes have been devised to control marketing and advertising practices. These laws now help to ensure that consumers are not coerced into buying products they do not want and are not otherwise deceived by sellers. Statutory cooling-off periods are an example of a legislative means of empowering consumers who are subjected to high-pressure sales techniques in their homes (see Goldring et al. 1998: 270–306).

In Australia, federal and state consumer protection laws apply to transactions in which Australian citizens or corporations are involved (see Goldring et al. 1998). The *Trade Practices Act* has provisions on consumer protection in Part V which proscribe various unfair practices and specify product safety standards and the operation of conditions and warranties in contracts.

The *Trade Practices Act* is, however, generally silent on whether its provisions apply to conduct carried out electronically, though the breadth of its controls would, arguably, apply to all online activities carried out between corporations and consumers. Most of the consumer protection provisions of the Act specifically apply to conduct that 'involves the use of postal, telegraphic or telephonic services' (s. 6(3)), which would seem to exclude the Internet and e-mail, which are not specifically 'telephonic'. This question has yet to be judicially determined in Australia, though the ACCC (1997a) takes the view that advertising on the Internet comes within the provisions of the *Trade Practices Act*.

Although Australia's consumer protection laws would apply to contracts for the purchase of goods and services entered into with merchants who have advertised on the Internet, the imposition of liability may be difficult and costly where overseas corporations are involved. Most laws apply only to transactions carried out between Australian citizens and corporations within Australia. The cost, inconvenience and logistics of cross-border legal proceedings make the imposition of liability on manufacturers, distributors and merchants outside Australia unviable for most consumers.

In the United States the FTC has a broad enforcement and consumer protection role. It is able to prosecute unfair or deceptive practices in or affecting commerce, including those that take place online (Starek and Rozell 1997). The *Telemarketing and Consumer Fraud and Abuse Prevention Act* 1994 extended the powers of the FTC to deal with those who undertake a pattern of unsolicited calls which a reasonable consumer would consider coercive or abusive of the consumer's right to privacy. Telemarketers must now identify who they are, say who they represent, and indicate the nature of their call. A maximum penalty of ten years' imprisonment now applies to wire fraud carried out in the course of telemarketing.

A number of other statutes are relevant to online deceptive practices. These include the *Fair Credit Billing Act* 1994 (15 USC 1666), the *Mail or Telephone Order Merchandise Rule* 1997 (16 CFR 435) and the *Electronic Fund Transfer Act* 1994 (15 USC 1693), each of which have been used in FTC prosecutions of online business practices (Starek and Rozell 1997).

In addition to these statutory provisions which seek to prohibit objectionable practices, it has been suggested that laws be enacted which would alter the behaviour of merchants by reallocating liability for online transactions that fail because of misleading or deceptive practices. Such a market-based solution is being considered by the OECD, which is exploring the possibility of some international agreement being reached that would enable chargebacks to take place by credit card issuers where an online transaction has failed for some reason. This initiative has the support of the Office of Fair Trading in Britain, which has similar procedures in place for consumer credit transactions (see s. 75 *Consumer Credit Act* [Eng.]; Bridgeman 1997).

Soft Regulation

In view of the practical difficulties associated with relying on legislative regulatory approaches to control misleading and deceptive online conduct, a number of industry groups have established self-regulatory mechanisms which have devised their own standards and codes of practice (see Chapter 11). These were originally created to deal with non-electronic forms of advertising and marketing, but are now being extended to deal with conduct in the digital world.

Content regulation

One of the primary strategies that seeks to prevent misleading and deceptive material from being disseminated electronically is the regulation of online content. This can be done through the use of screening software or through measures that require ISPs to monitor the material being publicized on their networks. Recent proposals have also entailed an element of hard regulation through the criminalization of content deemed unsuitable which has been publicly disseminated. Although the use of screening software has been widely advocated to control access to obscene and objectionable materials (see Grabosky and Smith 1998), its use in the control of misleading and deceptive advertising may be more difficult. Often the deception could not be discerned from an image or description of the product in question, and it might be impossible to differentiate between a legitimate advertisement and one that contained some misleading content, merely on the basis of the words or images used. It would also be difficult to influence those who provide content from overseas jurisdictions without some international agreement on what advertising material was considered acceptable by all.

Certification and endorsement services

As an alternative to the use of prohibitory schemes that seek to identify objectionable content and prevent users from gaining access to it, a number of certification and endorsement services have been established which provide users with information on the reliability and acceptability of online material. Users are then free to decide whether or not they wish to make use of the material in question.

In the United States, the Council of Better Business Bureaus performs a certification service in which Internet business sites are given a form of approval. Sites that display the authorized and encrypted seal of approval agree to abide by the Council's truth-in-advertising standards and to adopt its dispute resolution procedures. Members of approved Internet associations are able to display the fact of their membership and consumers are able to check to see if organizations do in fact have membership.

The WebTrust program, which was developed by the American Institute of Certified Public Accountants (AICPA 2000), certifies Internet sites that demonstrate sound online business practices after having undergone an extensive auditing procedure. The audit, which varies in cost depending on the complexity of the business and the site, includes checking the site's security measures, privacy practices and transaction-processing systems. The service is available from any WebTrust-licensed certified professional accountant or accounting company.

Since the AICPA began the WebTrust program, some 1500 CPAs and 75 accounting companies have been qualified as able to perform WebTrust audits (Tweney 1998). To date, only a small number of sites have successfully undergone the audit process, permitting them to display the WebTrust seal on their site. Like other third-party certification programs, WebTrust depends for its success on widespread acceptance by online merchants and users, which it is hoped will be achieved in time.

Certification and endorsement services have two main benefits. First, consumers are able to rely on the fact of a merchant being certified in order to have some measure of confidence in the trustworthiness of that merchant and in the availability of redress mechanisms if problems arise. Second, financial institutions involved in providing payment facilities could be encouraged to deal only with certified merchants who have agreed to comply with a code of conduct that meets certain minimum standards. This would provide a powerful industry-based inducement for merchants to undergo certification and to act responsibly and in conformity with established codes of practice.

One of the main problems with endorsement and certification is the proliferation of services and the determination of appropriate standards. Already, some twenty so-called 'Webseals' are in circulation in Australia, with the government providing a comparative table that sets out their various attributes (Cook 1999; DCTIA 2000a). Determining acceptable standards and publicizing these will represent a major challenge for the future.

Information intermediaries

In Australia, the ACCC (1997b) has also raised the idea of using information intermediaries to provide information about online merchants and the procedures involved in conducting business online, similar to the kinds of information that insurance or mortgage brokers provide. Several private enterprises, including the Australian Consumers Association, have set up an independent advice service on loans and mortgages offered online by financial institutions, while various consumer subscription services publish independently conducted evaluations of products offered online.

Preventive Strategies

In addition to the use of both hard and soft regulatory approaches, much can be achieved by way of self-help strategies which aim to alert users to the presence of misleading and deceptive practices. Such preventive action may take the form of regular surveillance of the Internet in order to locate objectionable and illegal practices (see Chapter 11), providing educational material warning users of dangerous schemes, and the use of authentication technologies to permit individuals to know with certainty who they are dealing with in the online world.

Most regulatory agencies throughout the world provide information on misleading and deceptive practices in paper form and electronically through Websites. One of the most comprehensive booklets directed at Australian consumers is *The Little Black Book of Scams* published by the Ministerial Council on Consumer Affairs (1999). In electronic form, the ACCC's website gives advice on pyramid selling schemes, business opportunity schemes, and phoney prizes and lotteries. Examples of popular deceptive practices are listed along with the legal penalties that apply to those who run or participate in such activities. In addition, and in order to enhance consumer confidence in the Internet, the Australian Government has produced a series of fact sheets that inform consumers about the risks of shopping online, and certain other issues such as paying tax and duty and privacy issues (DCTIA 2000b). Similar advice is available in Britain at the Office of Fair Trading (2000) and in the United States at the Federal Trade Commission (FTC 2000).

Consumer groups also represent a good source of trusted information for consumers. Groups such as the Australian Consumers Association, the Consumers Union of the United States and the Great Britain Consumers' Association conduct their own testing of products and services and publicize the results through subscriber-based magazines such as *Choice* (Australia), *Which* (United Kingdom) and *Consumer Reports* (United States). Although consumer organizations already provide consumer information by various means, including the Internet, perhaps the role of consumer groups here could be increased.

Finally, in order to solve the problem of consumers and merchants adopting false identities for fraudulent purposes when carrying out online transactions, electronic user authentication systems are starting to be employed. Public key systems that make use of encrypted data transmissions are one way of helping to ensure that both consumers and merchants are confident of the identity of the person they are dealing with. Such technologies would not prevent individuals from illegally getting access to private cryptographic keys by stealing tokens that hold keys or by presenting fabricated documentation in order to obtain key pairs fraudulently (see Office of Government Information Technology 1998). But they do represent a much more secure way of conducting online transactions than simply by trusting material that is displayed on the Internet and hoping that it will be secure.

Conclusion

Although some may question their effectiveness, online advertising is already subject to a variety of laws and other regulatory controls. Those who engage in misleading and deceptive practices invariably infringe local laws in the jurisdiction in which they reside or in which their material is read, or sometimes both. This often provides sufficient jurisdictional basis for the commencement of legal proceedings. The last 30 years have seen continual improvements in consumer protection legislation and dispute resolution procedures, and many online activities fall within the scope of these initiatives. International consumer protection agencies have also been created such as the International Marketing Supervision Network (2000), a membership organization that consists of the consumer protection law enforcement authorities of over two dozen countries. Its mandate is to share information about cross-border commercial activities that may affect consumer interests and to encourage international cooperation among law enforcement agencies.

Unfortunately, the remedies that are available to those who have been deceived electronically are often practically unavailable as they would require offenders to be extradited from other places or victims to take cross-border legal proceedings. Such action is invariably beyond the means of most individuals and costs far more than the amount lost in most consumer frauds.

Legal enforcement proceedings in the form of class actions against large corporations can sometimes be taken on behalf of groups of consumers who have suffered loss. Although these are sometimes slow and costly, victims are empowered through the weight of numbers and compensation can occasionally be made. But the perpetrators of many online scams are often not large corporations. They are able to close down their operations quickly and easily, move assets to secure locations, and use digital technologies to conceal their identities and disguise evidence. In such cases there is little likelihood of success whether civil or criminal proceedings are taken.

Consumers who transact business online need to be made aware of the risks they face and be informed about the nature of misleading and deceptive practices that are present. Already substantial amounts of information of this kind are available. The challenge lies in ensuring that consumers are made aware of its existence. In this regard, certification and notification systems, which allow consumers to identify readily businesses that have been found to be trustworthy, seem to provide the best option. Technology needs to be developed, however, to ensure that certification services are themselves unable to be manipulated. There could, for example, develop a trade in fraudulently acquired certificates of propriety which illegitimate businesses could attach to their website. Fraud relating to the process of certification might also develop in the future, as might the use of 'phoenix businesses' which re-establish themselves immediately after they have been closed down because of improper practices.

As with the other topics being considered in this book, misleading advertising and deceptive practices that take place electronically cannot be controlled by recourse to legal avenues alone. Although regulation by law has an important role to play, particularly in establishing the boundaries of acceptable conduct, commercial and business solutions, or what Lessig (1999) calls 'regulation by code', will often provide online consumers with greater protection than seeking to harness the jurisdiction of the courts across the globe.

CHAPTER 8

Intellectual Property in Cyberspace

The misappropriation of intellectual property is hardly unique to the digital age. The law of copyright has an interesting origin, sprouting as it did in the eighteenth century from existing efforts to control sedition. In responding to the challenge of an earlier communications revolution – the one brought about by the invention of the printing press – authorities in England, concerned over the threat to the security of the state posed by uncontrolled printing, exercised censorship through the licensing of printers. What we now would call 'content regulation' was achieved by chartering stationers by letters patent, and authorizing them or their representatives to search out and destroy illicit books and unauthorized printing equipment. From the licensing of books before publication evolved the system of copyright as we know it today (Kaplan 1967). The first legislative manifestation, the *Act of Anne* (1709), was entitled *An Act for the Encouragement of Learning, by vesting the Copies of printed Books in the Authors or Purchasers of such Copies, during the Times therein mentioned.* Copyright protection was extended to engravings in 1734, with the encouragement of William Hogarth (Ricketson and Richardson 1998: 64).

Intellectual property laws have long existed to encourage the authorized dissemination of innovative endeavour. While such innovations were, until recently, generally embodied in material things, many now originate in or are transmuted into digital form. Digital form enables information, once located, to be copied and transmitted rapidly, with few transaction costs. For obvious reasons, this poses a threat to the right of authors to claim exclusive authorship of their material. Recent technological advances have galvanized both intermediaries and producer and consumer groups, as current behaviours are fundamentally challenged. Electronic commerce is increasingly focusing on the means of trading intellectual property rights electronically. Digital transactions in respect of intellectual property objects, particularly copyright objects, present both opportunities and threats, as information flows are freed up and new industry structures begin to emerge.

This chapter examines the effect of digital technology on copyright, through the lens of the problems experienced in this area. Copyright has suffered the greatest impact from the information revolution. Copyright law has also felt the full force of globalization such that, despite local differences in definition and applications, the problems facing that area of law are universal. Globalization itself brings difficulties for enforcement of copyright since distribution of works crosses international boundaries and may occur simultaneously in several distinct jurisdictions.

The issue that increasingly arises is whether the expansion in technology warrants expansion in the law to match, or, indeed, whether the law has reached an impasse and technology will deliver the only viable solution. This chapter will consider the example of MP3 litigation (see page 137) to demonstrate the manner in which the law has been unable to adopt a truly 'technology-neutral' position to encompass all the actions that might be taken by 'copiers' in relation to that technology. It also argues that firms have attempted, to a degree, to replace reliance on breach-of-copyright actions with technological protections. Ironically, whenever a technological protection for a digital copyright object is devised, the work on subverting that protection begins. This makes for a constant spiralling between protection and circumvention. The law has tried to break this spiral by making the possession or creation of circumvention devices unlawful, but this response also has the potential to collaterally affect the public interest. This chapter examines such implications of technological advancement on copyright from legal, social and political economic perspectives.

Technological Change

The nature of change: from atoms to bits

Before the late 1980s, most intellectual property existed in some physical object, such as a book produced on paper. Given a tangible form to a work, copying was limited mainly to reproducing the object itself, and thereby its underlying technology. Generally speaking, because objects were captured in physical form, they could only be enjoyed by one user at a time. Most media could be replicated, subject to transaction costs. Because artists and publishers felt that ease of copying meant a threat to profits, public outcry accompanied almost every readily available device that was capable of reducing the transaction costs: witness the debate relating to photocopiers, blank audio tapes and video cassette recorders.

The debate over these articles centred on the increased ease of bulk reproduction – an effective lessening of the artificial scarcity imposed by intellectual property rights. Responses to this ranged from attempts to levy a surcharge on blank tapes, to endeavours to find contributory breach of copyright by the mere existence of the ability to use the technology for circumvention. Fundamentally, however, the economics of distribution remained largely unaltered. Only unsophisticated business models were required to ensure returns from licensing and assignment: for instance, a book or a compact disc could be purchased, giving a consumer the right to private use of it.

The ability to deliver materials in electronic form causes a significant shift from the traditional distribution models. The mode of production will in many cases be synonymous with distribution: for example, music can be produced, as David Bowie has done, in digital form and distributed as an MP3 or WAV (sound file formats) file. The significantly reduced time lapse from generation to distribution has advantages in that it decreases bottlenecks in the process. But it also generates the possibility of rapid copying by unauthorized persons.

Actions to mitigate the problem of unauthorized reproduction can take many forms, but organizations with vested interests in the area of physical distribution have a particular incentive to try to restrict the means of production and distribution to those to which they are accustomed. In relation to copyrighted objects, the approach has been twofold: to call for higher levels of protection and longer duration of protection, and to adopt technological measures to make the digital environment mimic the physical environment. Early attempts at technical measures include so-called 'dongles' (hardware protections attached to computer serial ports) and complicated passwords printed on special paper that could not be photocopied. Such measures have now been superseded by software measures such as encryption (the scrambling of data) and steganography (the art of hiding data within other data, e.g. digital watermarking).

Types of copyright transactions

In order to understand how digital technology has altered the world of copyright, it is important to understand the kinds of transactions that can transpire in relation to copyrighted objects. These can be quite variable because, given their nature as choses in action, copyrights are essentially a bundle of rights that can only be enforced by law. The typically discrete nature of the rights means that there are numerous transactions that can exist in respect of any given copyright property object.

Exclusive rights that exist include (depending on the jurisdiction) moral rights (Stefik 1997) and transport rights such as to copy, transfer and to lend, in addition to render rights such as to display, to print, and to play. Such rights may also include exclusive derivative rights, including to extract, embed, modify or adapt, as well as backup rights, to enable the licensee to protect himself against accidental loss of or damage to the licensed copy.

These rights may be either assigned or licensed. Assignment is effectively transfer of the entire bundle of rights in respect of an intellectual property object. A requirement of most copyright assignments is that they satisfy the statute of frauds provisions (which require certain contracts to be in writing and signed). Licensing is a more complicated matter. It involves giving away only part of a right, so that various interests are able to be created.

Licensing interests include exclusive licences (a licence for the entire bundle of rights contained within a limited geographic location or jurisdiction), period subscriptions (a licence for the whole bundle of rights for a limited time), and prepaid subscriptions (a licence relating to a designated number of intellectual property objects limited to a nominated set, and possibly restricted to a period of time). More specific licensing interests include site licences (a licence is granted in respect of each additional user or object satisfying a set of conditions such as being a subscriber to a database within a nominated domain) and sponsor-funded licences (a licence is granted with the proviso that consideration is to pass to a third party not privy to the agreement).

Other common licensing arrangements may consist of shareware licences (a licence is given free to the world by requesting that users forward a contribution), crippleware licences (a licence is given free to the world but for a very limited time, such as 30 days or for a limited use, such as three levels of a game), and freeware licences (a licence is given free to the world but possibly containing conditions such as requiring that the object not be used for commercial purposes).

These licences all represent the kinds of terms that can be inserted into contracts or licensing arrangements. Recently such arrangements have increasingly been traded online.

Online trading of copyrights

Detecting unauthorized reproductions of copyrighted works is easier for tangible objects (books, paintings, business documents) than it is for intangibles (digital representations), as there are likely to be physical differences between originals and copies, or between authorized and unauthorized copies. A significant challenge to online trading of copyrighted works in digital form is how unauthorized reproductions can be detected. Given the nature of digital storage of data, a copy can be perfectly identical to its original. This changes significantly in the face of copyright objects that are valuable in their intangible form. Large-volume distributed systems (such as the Internet, which can distribute massive amounts of information very widely) are of such a nature that a user or purchaser may be able to determine provenance precisely. This, combined with the broad range of assignment and licensing possibilities outlined above, has given rise to a desire for rights management systems.

> The very properties that make the Internet attractive as a distribution medium – ease of manipulating information in electronic form – also appear to make these protections intractable. Addressing this dichotomy requires a paradigm shift in computer architecture to introduce the concept of a 'secure processing'

> environment in which protected information can be manipulated without being subject to external tampering or disclosure. A prerequisite to such an environment is a cryptographically protected 'container' for seamlessly packaging information and controls that enforce information use rights. (Sibert, Bernstein and Van Wie 1996)

Sophisticated rights management systems such as these combine encryption with rights management languages and steganographic measures (Johnson 2000). Simply put, this means that a combination of scrambling the data so that only authorized persons may access it (cryptography) and putting hidden cues in the data so that if people make unauthorized use of it that use may be evidenced (steganography) can work to secure online copyright and transactions in online copyright.

An instance of such an approach is the alliance between IBM and Xerox, who in 1996 announced a joint venture to wed IBM's 'Cryptolope' containers to Xerox's Digital Property Rights Language (DPRL) (IBM 1996). 'Cryptolope' – 'cryptography' plus 'envelope' – provides a cryptographic shell which is used to house digital content and facilitate payment. For example, an image could be put in a cryptolope, which could be sent on to another person who would then have to pay to receive the key to unlock it. This would have demanded constant human interaction, which could be time-consuming. This is where Xerox's DPRL helps by providing a language that allows content providers to electronically designate actions sanctioned by end users with respect to a specific intellectual property object. A similar rights management system concerns erights (Erights 2000) and InterTrust Technologies Corporation's 'DigiBox'. The advantage of such systems is that the terms of a licence are able to be unambiguously phrased using a series of algorithms that would allow automatic ordering of copyright objects without the need for human intervention.

Licences could be prepaid by other mechanisms. At present the creators of such systems are competing to become the de facto standard in the area. Such proprietary systems are by no means a panacea. The best proprietary standard is not necessarily the one that will become the dominant standard, since pre-existing market power can prove very influential. The standards race is of special concern because of the possibility that fair use and statutory licensing will be severely curtailed as a result of such systems. If rights management systems do become the chief way of distributing copyright objects online, the object would have to be made accessible by some other means for an opportunity to engage in fair dealing to emerge. Electronic publishing effectively, therefore, extends the power of a copyright-holder. If copyrights are surrounded by high walls and virtual razor wire, fair use will be a moot point for users and for information repositories such as schools, universities and researchers since the mode of disseminating the information can limit access to an object to paying customers. Already, suggestions have arisen as to how this problem could be overcome. One is the concept of the digital library, from which objects could be 'lent', such objects becoming unusable after the expiry of the term of the 'loan' (Stefik 1997).

Rights management systems, even moderated by actions such as a digital library, do, however, present their own problems, for example failure to provide for anonymity and pseudonymity (Clarke and Dempsey 1999). The anonymity or pseudonymity could conceivably be on either side of the trade; authors have long embraced both, and cash purchases have effectively been anonymous. Clarke and Dempsey (1998) observe: 'A relatively small proportion of access has been associated with an authenticated identity; and that has generally been where the material was being adapted or incorporated into another work'.

Typically, firms dismiss these issues by the standard argument that those with nothing to hide do not fear identification. Some would no doubt argue that the advantage of not allowing anonymity or effective pseudonymity is that this would establish relatively accurate audit trails, thereby simplifying the tracking of illicit activity. The argument against this is that intellectual property rights are not intended to be the means of facilitating censorship. Free speech and censorship matters are in practice determined within a nation or culture. The argument that 'innocent' people have nothing to fear is heavily dependent on the social mores of the place and time. In an era of globalization of electronic commerce this is a dangerous notion to embrace: for instance, materials allowable or mainstream in the United States, Canada or Australia will offend certain laws that are (to Western perceptions) antithetical to free speech or liberty – consider the implication, even now, of a Chinese national purchasing a Falungong religious text online. Even on a conservative argument, therefore, anonymity and pseudonymity can be seen as important to modern electronic commerce as supporting democratic environments by removing fear of recrimination by authorities such as government, employers, or other powerful vested interests.

Other problems in electronic rights management systems also relate to vested interests. At present the role of publishers and distributors is changing online – middlemen now provide services such as selection of objects. Part of the advantage of the Internet is that the middleman need not exist in every trade. Some rights management systems (e.g. DigiMark), however, presume that publishers and distributors will continue to exist in their current form online, supplementing clearing houses and interposing in every trade. If producers of rights management systems presume that a middleman must exist in their schema of a trade, then this will occur, as the nature of the computer system will, to a degree, drive the trading options. Samuelson observed that:

> The biggest challenge that cyberspace poses for authors and publishers is not how to strengthen copyright law, but how to reinvent their business models so that they figure out how to provide content that will interest potential customers on terms that these consumers find acceptable ... In this new environment, copyright law will likely play a useful, if somewhat minor, role. (1996)

All these issues could be resolved through thoughtful international standardization. Attempts at this have been made by initiatives such as Imprimatur and Propagate, the latter of which developed a generally applicable model in a formal modelling language. An advantage that these ventures possessed over other rights management systems was the consideration of all stakeholders, rather than merely scheme sponsors. Unfortunately, this advantage may have contributed to the lack of commercial interest shown in such projects. As mentioned above, another standards battle is brewing over rights management systems. Until such time as a standard appears a clear winner, protection of digital intellectual property will largely remain the province of bare protection devices such as either cryptography or steganography rather than rights management systems proper.

Technological protection measures

Where matters are not subject to a rights management system, other modes of protection are adopted. This is typically limited to encryption for transit and storage and steganography (such as digital watermarking and digital fingerprinting) for tracing. The objectives of such technology are twofold: to prevent unauthorized copying (encryption) and to ensure a trail of evidence if unauthorized copying does occur (steganography). Watermarking (including fingerprinting) means inserting identifying data (such as copyright information) into other data, such as text, images, and even moving images. Watermarks can be perceptible to the naked eye, imperceptible, or a combination of both. A perceptible watermark will typically be designed to prick the conscience of a user by clearly identifying that object as proprietary. Imperceptible watermarks are usually generated by using an algorithm to insert the data into an object, marking that object as proprietary. Where such an algorithm is adopted, if the object is copied without satisfying certain conditions, the watermark will act as evidence against the copier. Imperceptible watermarks may also contain information that allows other programs, sometimes called 'bots', to track them where the copy appears on a Web page. One example of this is the Digimarc system, which uses a digital ID (Digimarc ID) in conjunction with a watermark embedder (Digimarc Embedder) and allows an optional subscription to a bot called 'MarcSpider' to track the images across the web and to report and archive where the image was found and a date range of hits (Digimarc 2000). Watermarks can also be used to direct data to effectively self-destruct if a number of stipulated conditions are not met. It is also possible to make watermarks behave in the same way as dye-packs in banknotes, but the adoption of such a program would present legal problems because it would effectively work as a virus. A sensible middle ground is the adoption of both a perceptible and an imperceptible watermark, so that users can know that the object is proprietary – if the user makes an unauthorized reproduction regardless of this information, they can (to a degree) be traced.

Digital watermarking is not without problems – like any lock, a legitimate key is not the only means of access. Watermarks that are perceptible can be removed by various measures such as resizing and rotation. Imperceptible watermarks are vulnerable to distortions that render their embedded data unreadable. Johnson, however suggests that tracing can be carried out by relying on salient features of an object, much in the same manner as individuals recognize faces (Duric et al. 1999). Watermarking standards are currently being pursued by both firms and academics. In February 1999, Hitachi, IBM, NEC, Pioneer Electronic, and Sony joined together to form the 'Galaxy group' to develop a unified standard for sending copy-protected data. Some specific initiatives are occurring in the area of music. An example of the application of watermarking technology lies in the proposed 'solutions' to the MP3 issue.

MP3 is the Moving Pictures Expert Group (MPEG) Audio Layer Three: basically a compression algorithm that is standardized as ISO-MPEG Audio Layer-3 (IS11172-3 and IS13818-3). This kind of algorithm is called a Codec (COder/DECoder). Many MPEG standards have been devised for different levels of encoding and data rates (Audio and Video 2000). Depending on the music that is being copied, the quality ranges from nearly equivalent to compact disc quality, to noticeably worse. The reason for a reduction in quality lies in the way in which the data are optimized to account for human hearing. Parts of the data (high and low frequencies) that lie on the edge of perception are removed. This causes some sound degradation.

Regardless of the reduction in sound quality, the advantage of MP3 is that it enables data (representing music) to travel across the Internet consuming relatively low bandwidth with rapid transmission speeds. Sound recordings can be delivered either by 'streaming' (like a broadcast or simulcast) or by merely saving a sound file (more like a CD). This naturally makes MP3 and its ilk a tempting means of reproducing artists' work, with or without their authority. According to Intersect, more than 2600 sites are currently involved in the distribution of pirated audio CDs. Illegal music downloads occur, depriving artists and publishers of royalties paid on sales of copyrighted materials.

Record companies were slow to perceive any issues arising from the creation of such a standard. In 1987 chip speeds were relatively slow, but the nature of desktop computing was such that rapid advances in chip speeds were foreseen. MP3 did not really become viable for the desktop user until the late 1990s when chip speeds had reached levels such as 300 MHz. When MP3 became an issue for the record companies, there was an outcry against the standard, including some calls to outlaw its use (forgetting, of course, that it formed part of a standard for streaming audio and video on the Internet). Interestingly, the proliferation of CD burners has not caused the degree of commotion one might expect. The large record producers and a few interested firms and individuals began considering ways of limiting MP3 such that it could not be used for unauthorized reproduction. Involved in this field are the Recording Industry Association of America's (RIAA) Secure Digital Music Initiative (SDMI); the RIAA are involved with InterTrust in an attempt to create a standard. Effectively competing against them are a number of firms such as Liquid Audio, who, in concert with Solana Technology, have developed a system that automatically adds digital watermarks to music purchased online (Minidisc 2000).

Regardless of the firm, a digital watermark in relation to a sound recording is basically data embedded into a file that contains copyright information such as International Standard Recording Code, user identification and other rights management data. When a copy of the file is made, these data are also copied. Depending on the method of imprinting the watermark, it can be made difficult to remove without seriously damaging the original file. The SDMI are currently developing a standard that protects sound recordings (and circumvents players such as the Rio) by only enabling the data to play on certified players – that is, players containing a chip that recognizes the interrogation encapsulated in the sound file. This form of protection is similar to that currently used to generate zones in relation to digital video disc (DVD) players, and is apt to be confronted with similar problems (recently, algorithm to make the DVD players work anywhere was released onto the Internet). If several such standards arise, it may cost more to play sound recordings owned by different record companies. Circumvention of such devices could be possible via the manufacture of 'all players' chips by third parties who have reverse-engineered the chips and incorporated their functionality.

Law and Economics

From the above discussion it can be seen that the issue of technological change and copyrights is a vexed one because of the nature of the heavily competing interests of the stakeholders. The issue of intellectual property online is polarized: either intellectual property is protected with the same absolute rights afforded to physical property, or information wants to be (and should be) free (Barlow 1994). The continuing saga of MP3s reveals not only the degree to which the law is perplexed and vexed by new technologies, but also the threats that new technologies have posed to intermediaries such as record companies and book publishers. The democratization of the Internet has led to 'disintermediation' – an excising of the middleman. The challenge this has posed to traditional intermediaries has been met with a flurry of litigation.

This is particularly exemplified by the conduct of the cases relating to MP3 and similar online music provision. The initial lawsuit relating to MP3 in the United States was that of *RIAA* v *Diamond Multimedia Systems Inc.* (USDC Central Dist. Cal., Case No. 98-8247 and on Appeal 180 F3d 1072). In that case, while RIAA did not directly allege copyright infringement, it did attempt to use the US *Audio Home Recording Rights Act* 1992 (17 USC ss.

1001–10) to effectively prevent the manufacture and distribution of Diamond's 'Rio' MP3 player. The grounds were that the Rio player constituted a 'digital audio recording device' under s.1002 of that Act and that it was consequently obliged to integrate a system for preventing the making of copies. This case failed on two main grounds. The first was that computer hard drives (to which a Rio player would be attached) do not constitute a 'digital audio recording device' under s.1001(3), which defines such devices as being 'designed or marketed for the *primary purpose* [emphasis added] of making a digital audio copied recording for personal use'. The second failure was that data contained on a computer's hard drive could not constitute a 'digital music recording' under that Act, as under s.1001 it expressly excluded 'a material object in which one or more computer programs are fixed'.

This case was followed in April 2000 by *RIAA* v *MP3.com Inc.* (Unreported). Rather than directly attacking the MP3 format, RIAA sought action against two services offered on the My.MP3.com service on the MP3.com website: 'Instant Listening' and 'Beam-It'. The former allowed the user who purchased a CD from an online retailer affiliated with MP3.com to elect to have the album placed on the user's account or 'locker' immediately. The latter, Beam-It, enabled users to load a CD onto their PC; MP3.com would then check the fingerprint on the CD to verify its authenticity. Once the CD had been confirmed, the user's 'locker' would be able to access that music (from a library of pre-generated MP3s) by the application of a password. My.MP3.com argued that the activities were similar to 'time shifting' as described in *Sony Corporation of America* v *Universal City Studios Inc* (464 US 417, 104 S.Ct 774). In the Sony case, 'time shifting', meaning taping a television program using a video cassette recorder for later private viewing and then erasing the work, was held not to infringe copyright. My.MP3.com argued by analogy that a 'space shift' enabled subscribers to the service to avoid the need to take their entire CD collection whenever and wherever they travelled. The court made findings of fact that MP3.com had copied the CDs, and that this constituted a breach of the copyright-holder's exclusive right to license sound recordings for reproduction. MP3.com's defence of fair use was rejected, as the court found that the purpose behind Instant Listening and Beam-It was commercial, by virtue of the revenue gained from advertising.

As MP3 technology has enjoyed increasing success, the manner in which copyright infringements can be avoided has also expanded. An ongoing foe of the RIAA is a firm called Napster Inc. This firm provides two related services in software called 'Musicshare', comprised of server software and a search engine. The server software enables any compatible computer to 'serve' MP3s – it effectively catalogues MP3s sitting on the computer's hard drive, setting up a 'library' that can be searched by a central database. When the user 'signs on' to Napster, the database is updated. If a user wishes to copy a recording, a connection is made between the client (the computer of the user who wants the recording) and the server (the user who is providing the recording) and the recording is automatically retrieved by the software.

Napster argued that it was entitled to take advantage of the *Digital Millennium Copyright Act* 1998 (17 USC s 512(a)), which provides a so-called 'safe harbour' for service providers. On application for summary adjudication of this issue, Napster's position was held to be different to a 'service provider' on the grounds that the MP3 files were not actually transmitted via their 'system'; rather, the recordings were passed between the users such that the only connection to Napster was the actual address of the various machines containing the recordings. Although Napster failed on s.512(a), they were held to be protected by s.512(d), which is designed to protect online search engines from breach of copyright actions provided that they have no actual knowledge of infringing material, that they are not aware of apparent infringing activity, that when made aware or when notified of infringing activity they disable or remove access, and that they do not receive a financial benefit directly related to the infringement.

In accordance with this provision, Napster has blocked from the search engine component the music of both Metallica and Dr Dre, but such blocking is of limited effect as it blocks user passwords from logging onto the search engine (other search engines can be used). RIAA remain in dispute as to the exact nature of the operation of the search engine component and whether this ought to enable Napster to avail itself of the imprimatur s.512(d). On 27 July 2000, the RIAA won a preliminary injunction in the US Federal Court preventing Napster Inc. from continuing its activities (AAP, *The Age* 28 July 2000: 3). The litigation surrounding the use of this technology will undoubtedly continue for some time.

Disintermediation and intermediaries' responses such as those outlined above are not necessarily efficient. In the long run, a more rational response is for the intermediaries to take advantage of their corporate goodwill and to change the nature of their role in a manner similar to that executed by Napster, from retailing physical goods to value adding by saving people the valuable asset of time. In economic terms, Napster has effected 'reintermediation' – it has put the middleman back into the transaction by taking advantage of the fact that searching a mass of online information imposes transaction costs on users. Because of its specialization and time advantage, Napster is better suited than the average MP3 consumer to locating MP3s. Throughout the information economy, disintermediation and reintermediation, specialization and exchange will occur whenever new technologies are adopted.

The example of MP3 only partly reveals the problems of copyright and new technology. Other areas exist that are not afflicted with the spectre of significant economic loss or gain. Instead, more 'traditional' problems arise – such as determining whether unapproved hyperlinking constitutes breach of copyright.

At first sight, hyperlinking is the objective of the World Wide Web, and as such placing materials on the Web can be construed as an implied licence. One particular technique, 'deep linking', however, has caused annoyance to some firms. Usually, when one visits a Web page, one arrives at a 'front page' (top page) which introduces the structure of the underlying pages. This is also the most common location for advertising. In the case of *Shetland Times* (Fact Net 1996), a news provider had been linking to *Shetland Times'* articles directly instead of linking to the front page. This meant that *Shetland Times'* readers could bypass the advertising. Although the equivalent of an injunction was issued in favour of *Shetland Times,* the case was eventually settled, but not before *Shetland Times* had suffered significant criticism for impinging on the freedoms of the Internet. Such outcry did not, however, accompany the case of *Washington Post Co.* v *TotalNews Inc.*, which involved 'framing'. With framing, a hyperlink to a new site, such as the *Washington Post*'s website, is presented within the frame, or website, of the 'jump-off' website, e.g. TotalNews. The URL (web address) displayed on the web browser may also still reflect the jump-off site. Effectively, framing can mislead users about which site they are viewing.

Regardless of the outcomes of such cases, changes in the technology, for better or worse, have inspired lobbyists who, in turn, have driven legislative agendas. Typically, such agendas involve increasing the level of protection. International business lobby groups such as the RIAA and the ITAA (1998) were successful in having the World Intellectual Property Organization negotiate a new Copyright Treaty (WIPO 1996), which under Article 11 requires countries to 'provide adequate legal protection and effective legal remedies' against 'the circumvention of effective technological measures that are used by authors in connection with the exercise of their rights'. Article 12 requires similar actions in relation to the removal or alteration of electronic rights management information. These requirements were met in the United States by the enactment of the *Digital Millennium Copyright Act* 1998, which cast the anti-circumvention net so broadly that otherwise legal and desirable activity could be criminalized (Garfinkel 1998). The impact of these provisions on fair use and fair dealing as well as statutory licensing remain to be ascertained; the danger is that such legislation, in conjunction with a steadily rising uptake of technological protection measures, may result in the demise of our current system of fair use, including its inherent provision for anonymity.

Some scenarios resulting from both increased copyright protection and increased uptake of technological protections may have significant implications for the unimpeded flow of information. At present, many publishers would have the public believe that the incentive for originating copyright objects and publishing is at risk. Proponents of public access to information, on the other hand, perceive there to be far greater risks for society if large corporations gain the enhanced copyright protections that they seek, use their monopoly power to impose contractual rather than copyright-based relationships, and/or apply technology in order to monitor and enforce. The major interests on whose behalf the WIPO treaty was prepared would appear to be winning their battle. They have supported legislation that allows activities that offend large copyright-ownership interests to be dealt with by the criminal courts (at the public cost) rather than the civil courts (at their own cost), and they have also won the argument to extend the duration of copyright by twenty years in the United States. The economic implications of such victories for heavy copyright protection may not be seen for some years, but the effect will be that fewer firms will control more information.

Conclusion

The opportunity for misappropriation of copyright objects did indeed increase during the early years of the Internet. Objects have been unlawfully translated into digital form and distributed at relatively low cost. At present, however, technologies are available to providers of copyright objects in digital form which allow tight control of their transmission through the use of encryption, and control of their replicability through the use of steganography. Using an analogy to the physical world, in the early days of the Internet, the homes were not locked, and people were robbed. Now, encryption and steganography provide copyright-holders with locks and keys. If a person were to be burgled as a result of having left their home unlocked, one might consider them imprudent for not having protected themselves adequately. The same might be argued for the use of protective technologies. At the same time, society must be careful to ensure that such locks and keys do not result in all property being controlled – access must be allowed to the intellectual commons just as it is to its physical equivalent. The arena of copyright presents a useful microcosm in which we will be able to view the efficacy of a legal pluralistic approach in the coming years.

CHAPTER 9

Industrial Espionage in the Digital Age

Espionage is said to be the world's second oldest profession, and the application of espionage to private economic information has a rich history. In addition to the immediate recipients and sources of the information in question, the winners and losers in the game of industrial espionage are often nations and civilizations. The Chinese monopoly on silk production was broken when Persian monks smuggled silkworms out of China on behalf of the Roman Emperor Justinian. Textile manufacturing in North America only became feasible when the technology of production, previously a closely guarded secret, was covertly transferred across the Atlantic from Britain. Understandably, governments will go to considerable lengths to protect their countries' trade secrets. Harris (1998: 10) describes how, in response to the eighteenth-century efforts of the French to obtain British technology, Britain enacted laws against the emigration of skilled workers and the export of machinery. Offenders were liable to confiscation of assets in Britain and could be prohibited from inheriting property under British law.

The motives for industrial espionage may not have changed much over the centuries, but some of the technologies certainly have. In the contemporary global economy, when a competitve edge may spell the difference between corporate success and disaster, private economic information is rapidly increasing in value worldwide. Although statistics in this area are unreliable, figures ranging between $50 billion and $240 billion have been posited as the amounts 'lost' in the United States through industrial espionage (Perry 1995). In 1999, more than half of the Fortune 1000 companies in the United States indicated that they had been targets for such activity (CNN, 9 March 1999).

Industrial espionage is, broadly speaking, misappropriation of an organization's private information by an external party not normally privy to that information. Where the misappropriation is conceived by government, the activity becomes economic espionage. If the information that the government covets is related to another nation's security, the activity is characterized as espionage per se. Any attempt to draw clear distinctions between industrial espionage, economic espionage and espionage per se is, however, doomed to failure, as the characterization of the activity is dependent on characteristics of information that are subject to change over time. A neat taxonomy of information intelligence has been adopted by Steele (1993):

> a. Open source or public information; within intelligence communities, this is known as open source intelligence or OSCINT. 'Grey literature', literature which is unclassified and not proprietary, but produced in limited quantities for limited purposes, is included as an element of OSCINT. Open (unclassified) electronic information, such as that available through the Internet and related file servers and newsgroups, is also included in OSCINT. The vast majority of scientific & technical intelligence is available through OSCINT, to include 600 scientific & technical journals that appear only in electronic form.
>
> b. Open proprietary information; discernible through open source investigation. This includes the reverse engineering of legitimately acquired products, and legally conducted 'competitor intelligence'. (Note: competitor intelligence is the globally accepted term for legal research efforts by businesses studying their competitors' products, organizations, and related matters.)
>
> c. Closed proprietary information [trade secrets]; available only through industrial espionage or clandestine and technical penetrations of regulatory agencies.
>
> d. Classified information [military and national secrets]; available only through clandestine human intelligence or technical (imagery or signals) intelligence.

Although industrial espionage is by no means a purely digital phenomenon, technological advances have increased both the opportunities and the motives for this activity. A firm's marginal chance of being subject to electronic industrial espionage increases with each additional computer that is attached to a telecommunications device, whether the link is to an internet or an intranet. Increases in industrial espionage being perpetrated electronically are readily foreseeable based on variables such as the nature of the firm, the kind of information stored on linked computers, and the degree of security and control exercised over the networked computers.

As a result of technological advances, legislatures have been faced with the delicate problem of determining the best balance of protection for private information. This is a particularly challenging task, as the variables affecting such a decision are not universal and rest on matters such as assumptions about the level of industrial espionage activity and whether a nation is a net importer or exporter of goods or services whose basis is private information. The issue is further complicated in that the public desires lower prices and for that reason values competition because it benefits them. Benefits derived from lower prices can be offset by the short-term damage that large-scale industrial espionage can cause to a nation's economy. The trade-off in such a situation is between long-term efficiency and short-term efficiency. The best 'mix' for each individual country will depend on variables such as the nature of their economic base (e.g. manufacturing or agricultural) and their legal notions of property.

How is Industrial Espionage Perpetrated Electronically?

Industrial espionage is an activity that particularly lends itself to telecommunications networks, as the spectrum of devices and protocols enables the dissemination of information at varying speeds in a relatively undetectable fashion. Types of information in high demand range from aeronautic systems, electronics components and energy materials through to less glamorous items such as lists of suppliers and clients. The nature of the perpetration will relate not only to the type of information, but also to the manner in which it is stored and the competence of the security.

In the absence of strict internal controls, the greatest opportunity for industrial espionage rests with employees and their associates, since they often have, or can get access to, the information. The methods used by such people to gain information are often not electronically sophisticated, other than in the elements required, if any, to circumvent security measures. A Kodak employee in the United States, for instance, was convicted of industrial espionage after attempting to send several large confidential files via e-mail to her sister, an employee of Xerox. The attempt was discovered when the sheer size of the files crashed the Kodak server, rather than through any sophisticated monitoring (Wired News 1998b). An example of a relatively sophisticated attempt was that planned against Ericsson Telecom in Sweden (Svensson 1997). A Cuban national was allegedly requested by his embassy to make a floor plan, including every room and its contents, which would be passed to another operative who had at the time gained employment as a cleaner in the firm. Attacks such as this, with the amount of planning required to place someone as an employee, are certain to be less common than those where employees are subverted. Organizations also often forget to have a healthy suspicion of other firms with whom they collaborate. Confidentiality agreements need to be carefully crafted; consultants, outsourcing and joint-venturers are often privy to huge volumes of sensitive information and are therefore the people ideally suited to provide data to competitors. The degree to which electronic means are adopted by such individuals in the collection of private information will vary inversely with the degree of access to that information. Activity in this domain is more likely to be opportunistic than premeditated.

Collection by competitive intelligence professionals is unlikely to involve any unlawful measures. Competitive intelligence is the process of monitoring the competitive environment; it is the creation of insights that help make better business decisions. Such professionals collect information from legal observation and open sources, and make educated estimations of activity because of their investigations. Intelligence professionals are no different to accountants and lawyers in that there are societies that provide accreditation and that impose a code of ethics. Ethical requirements include complying with the law, disclosing all relevant information, including one's identity and organization, before all interviews, and respecting all requests for confidentiality of information.

Although it is impossible to gather accurate statistics in the area, people hired to acquire industrially sensitive information by unlawful means are rarer than those hired to compile and interpret open source information. More often, acts effectively constituting espionage are executed by opportunists such as hackers. There are very few 'genius' hackers – the problem is that the code is distributed onto the Internet and many very ordinary hackers are able to make use of the technology. The implication of this for industrial or 'corporate' espionage is that the volume of hacking break-ins will increase and the opportunity for the misappropriation of private information will also increase. Depending on the accuracy, timeliness and nature of the information, demand will typically meet supply. An example of this is an event that occurred in Calgary, Alberta: industrial espionage thieves were arrested after they attempted to hack into Cybersurf, an IT firm. The IT central technology server had initially alerted the firm to the possibility of break-in and the firm followed this up by hiring private investigators. The subject of the attempted theft was the patent pending 'Virtual TI', a secure online transactions authorization system (Canadian Corporate News, 8 October 1996).

The proliferation of sites providing for online purchase of sophisticated surveillance equipment may itself indicate the potential for expansion of industrial espionage. Such sites advertise for sale items such as laser listening devices with advertisements along the following lines:

> We've received countless requests for these, so here they are! A laser listening device which requires no transmitter to be placed in your target's area! Simply focus the laser on any window of your target's home, office, etc. and hear what's going on. Once the laser hits the window, it instantly reflects back at the unit, which receives the returning laser signal, and gives you the audio from inside. The standard unit has a range of up to 50 meters, while the advanced unit features an additional lens increasing the range to 200 to 400 meters! (Espionage Unlimited 2000)

Hacking and telecommunications interception are not the only ways that industrial espionage occurs, but they are becoming an increasing prospect for those who would misappropriate such information. The challenge is to determine a form of regulating industrial espionage that does not remove the right to pursue competitive intelligence. Underlying such an issue is often the problem of defining information as property.

The Problem of Defining Private Information as Property

Information is an intangible subject to non-rivalrous consumption – that is, the possession of information by one individual does not preclude simultaneous possession by another. These key features mean that information lacks the characteristics necessary to fall within the crime of theft or larceny because the requirements of asportation (physically moving an object) and an intention to permanently deprive cannot be found. For this reason, in order to find 'theft' in information the law must characterize information in such a way as to limit its transmissibility. Ultimately the decision on the degree to which dissemination is to be restricted will rest on distinguishing between the notions of information as a common resource (private information) and ownership of information as a natural-law right (proprietary information). In this way it provides the foundation for the degree to which industrial espionage in the absence of other wrongdoing will be considered to be a civil or a criminal matter.

Relationships and Confidence

Where information is conceived as a resource, the degree of protection afforded it will be fairly low and will be characterized by the relationship between the 'owner' and the recipient of the information. The historical jurisdiction of equity is over those who have assumed a position of trust or confidence towards each other. Equity intervenes to protect that relationship (Sealy 1962). Jones (1970) argues persuasively that the entire basis of the action in breach of confidence lies in the duty of good faith. The protection afforded by breach of confidence is, on principle, limited to protecting confidees within that relationship. In such a conception the onus is on the confidor to make it clear to the recipient by words and conduct that the information is given in a mutual relationship of trust and confidence, and that the confidee is sufficiently aware of this to be taken to have submitted to the relationship (Finn 1984).

The test used is whether a reasonable man standing in the shoes of the recipient would have appreciated his position of confidence (*Coco* v *A.N. Clark [Engineering]* [1969] RPC 41 at 47–48 per Megarry J.); applied in *Half Court Tennis* v *Seymour* ([1980] 53 FLR 240 at 255 per Dunn J). No breach of duty occurs unless 'a confidence reposed has been abused' (*Smith Kline and French Laboratories [Australia] Ltd* v *Secretary, Department of Community Services and Health* [1991] 99 ALR 679 at 692 per Sheppard, Wilcox and Pincus JJ). That the duty is limited to the relationship of confidence is evident from the inability of other parties affected by the disclosure of the information to maintain an action. Also, the confidor may be found to have failed to take reasonable care and thereby published the information to the world. Confidential information is not to be confused with an idea *simpliciter*: information must be clearly identifiable, of potential commercial attractiveness and sufficiently well developed to be capable of realization (*Fraser* v *Thames Television* [1983] 2 All ER 101). In that case, the distinction between a mere idea and information was exemplified as a vague idea for a television show that was held to be insufficiently substantial to attract a duty of confidence, as much work would have to be done to bring the idea to realization. While the information must be more substantial than an abstract idea, it need not necessarily be novel or original. In *Talbot* v *General Television Corporation* ([1980] VR 225), Talbot had developed a concept for a television show about millionaires and had prepared written submissions, stories, and formats for negotiations with Channel Nine. Whereas there was no commitment on behalf of Channel Nine to take up the show, Talbot was successful in obtaining an injunction to prevent them from showing a similar program later. The court held that the concept was developed to the point of establishing a format in which it could be presented and that therefore it was more than a mere idea. The fact that there is little novelty in the concept of a show about millionaires did not preclude the information from having been provided in confidence.

Once information is made available to the public, the confidentiality is removed (some US judges have historically adopted this position; see *International News Service* v *Associated Press* [1918] 248 US 215 at 250). In *Malone* v *Metropolitan Police Commissioner* ([1972] Ch 244 at 376), Megarry V-C. held that a person who utters confidential information must accept the risk of any unknown person overhearing that is inherent in the circumstances of communication. These rules suggest that the jurisdiction is designed to protect fiduciary business relationships and not the confidential information per se. An exception may be the case of espionage, which would fall under the 'reprehensible means' jurisdiction of equity.

The authorities diverge in relation to the receipt by third parties of private information. In English law the approach appears to be that if the third party is unaware of the confidential nature, they are not bound by the equity even where they later become aware of that nature. By contrast, Australia appears to adopt a notion akin to the 'springboard' doctrine whereby if a third party becomes aware that information is confidential after acquisition, the equity binds them from the time of their cognizance of that fact. At Canadian law the springboard doctrine notionally applies, but the remedy of final injunction is not likely to be awarded because it is seen as contradictory to the patent system.

Common-law Copyright as an Improper Analogy

It is true to say that the focus on the abuse of relationship in English-law systems carries over to the criminal arena, where the class of crimes relating to information misappropriation is narrowly limited to breach of duties provisions such as those relating to the duties of company officers. In English jurisprudence the reason for regarding information as property of itself seems to have arisen from a narrow class of cases in which there was some pre-existing common-law copyright – particularly as to letters and unpublished manuscripts (*Pope* v *Curl* [1741] 2 Akt 342; *Gee* v *Pritchard* [1818] 2 Swans 402; *Lytton* v *Devey* [1884] 54 LJ Ch 293; *Abernathy* v *Hutchinson* [1824]

1 H &Tw 28 at 35–37; *Jeffreys* v *Boosey* [1854] 4 HLC 815; *In Re Dickens; Dickens* v *Hawksley* [1935] Ch 267). The protection granted to these 'common law rights of property' went far beyond that accorded to statutory copyright, even today. It is to copyright, and not some new jurisdiction, that judges speak when referring to confidential information as property in an English-law perspective (Meagher et al. 1992). The remedies of injunction followed as a natural consequence, but are not authority for any wider propositions in present law. The tail of the remedy, typified by injunction (Vaver 1979) is seen in many cases to wag the dog of duty (Finn 1979: 160). Apart from these special cases, the legal basis for the jurisdiction in English law is rather the equitable principle of good faith (*Seager* v *Copydex Ltd* [1967] 1 WLR 923 at 931 per Lord Denning MR). Turner (1962: 404) notes in contrast that the modern cases focusing on good faith provide only occasional dicta holding the innocent third party liable.

Natural-law rights and the US legislative approach

In complete contrast to characterizing the right by the relationship, the 'natural-law' thesis assumes that a man has a natural property right in his own ideas and that therefore the unauthorized use of information is theft (Lahore 1981). This thesis is based on seventeenth and eighteenth-century concepts of natural rights that were embodied in the legal systems of revolutionary countries like the United States and France. This natural-law notion of proprietary rights in information produces legal consequences unfamiliar to English-law systems, though such notions are gaining in popularity as a response to demands to protect wealth. This has resulted in great doctrinal confusion and has caused the jurisdiction to grow to the point where the scope of liability and remedies approach that of tortious interference with chattels (Birks 1985). The ultimate expression of information as property occurred in the United States, which went further than merely describing commercially useful ideas and information as 'proprietary information' by enacting the *Economic Espionage Act* 1996. In the Statements on Introduced Bills and Joint Resolutions, it was claimed that the bill addressed the 'systematic pilfering of our country's economic secrets by our trading partners which undermines our economic security. It would not be unfair to say that America has become a full-service shopping mall for foreign governments and companies who want to jump start their businesses with stolen trade secrets'.

The justification for the bill was that the money spent in acquiring and creating proprietary information was inadequately protected by existing laws. United States civil trade secrets laws were thought to possess insufficient deterrent power, while the *National Stolen Property Act* was thought to provide inadequate protection as some courts held that 'purely intellectual property' did not constitute the theft of 'goods, wares or merchandise'. Federal mail and wire fraud statutes prohibiting the obtaining of property by false pretences or representations were also relatively unsuccessful for protection purposes, because of the need to connect a 'scheme to defraud' with the use of the mail or wire transmissions to perpetrate the trade secret theft.

The provisions contained in the Act are cast quite broadly and have two effects: to make theft of trade secrets a crime; and to make economic espionage a separate crime. Section 1839 defines trade secrets as including all forms of information (regardless of the form in which it is stored) that might have economic value 'from not being generally known' to 'not being readily ascertainable through proper means', provided that the owner had taken 'reasonable measures' to keep the information secret. Theft, attempted theft, and conspiracy to commit theft or proprietary information from a US owner for the benefit of a foreign government or a corporation, institution, instrumentality or agent are criminalized under section 1831. The definition of theft under section 1831 effectively has two broadly cast limbs. The first prevents traditional types of theft: 'steals, or without authorization appropriates, takes, carries away, or conceals, or by fraud, artifice, or deception obtains such information'. The second reflects copyright-like protection without reflecting the corresponding public interest: 'without authorization copies, duplicates, sketches, draws, photographs, downloads, uploads, alters, destroys, photocopies, replicates, transmits, delivers, sends, mails, communicates, or conveys such information'. Other subsections relate to dealing in stolen goods, conspiracy and attempt.

Disposal of trade secrets is exhaustively defined, including the alteration, destruction, replication and communication of the purloined trade secret. The provision relating to receiving stolen information is broad, outlawing receiving, buying, or possessing a trade secret while knowing it to have been stolen or appropriated, obtained, or converted without authorization.

The actual economic espionage provision, section 1831, mirrors the trade secrets provisions but has a requirement of proving that the defendant intended or knew that the offence would benefit any foreign government, foreign instrumentality, or foreign agent. This definition appears deliberately broadly cast. By saying 'any' foreign government instead of 'a' foreign government, the burden on the prosecution is somewhat alleviated as the causal nexus between the theft and the benefit need not be as direct. To reflect the increased severity of the crime, it has higher penalties: fifteen years' maximum term as opposed to ten years for mere theft of trade secrets. For both theft of trade secrets and economic espionage there is provision for the victims to also receive restitution from funds recovered and the potential for import bans against goods created by the stolen technology. In addition to this, the Act provides the court with power to issue protective orders and to take appropriate measures including interlocutory appeals against decisions or orders authorizing disclosure of the information.

Part of the United States' decision specifically to criminalize economic espionage rather than relying on civil remedies was because a private suit against a foreign company or government 'often just goes nowhere'. The

Economic Espionage Act 1996 provides for extraterritorial jurisdiction where the victim or offender is a US citizen or where the offence was intended to have, or had, a direct or substantial effect in the United States.

Conflict with Dissemination of Information

Such an expression of protection of private information demands a consideration of the balance between public and private interest. Two economic reasons are generally forwarded for enforcement of confidences. First, the non-rivalrous consumption nature of information is said to remove incentive to produce information (Gurry 1984). Preservation of secrecy acts as an incentive to intellectual production by enabling its possessor to retain the competitive advantage that the secret confers, thereby affording the opportunity for a compensating premium for research and development costs.

Second, the law is said to give rise to an efficient framework for the operation of economic enterprise. By protecting secrets the law reduces the need for firms to adopt costly security measures. It also gives incentive for firms to efficiently use specialist contractors which they might otherwise avoid for the same security fears. The cost for the owner of technological information of contracting with other firms possessing complementary information or resources is said to be reduced, and wasteful security measures taken as self-help in the absence of property right protection are avoided (Kitch 1977). The latter consideration provides justification for equity to protect confidences within particular relationships.

There are evident short-term advantages for any firm of dealing confidently with employees and contractors. An effective monopoly in some 'information' provides a firm with the opportunity to take advantage of that information by treating it as an item that can be traded. The disadvantages, however, are less obvious. Information, according to Lamberton (1994), should not be conceptualized as discrete parcels or 'bits' of data but as an organic whole – a 'flow' throughout organizations. Conceiving information as a flow means that it can be understood to be contingent on other 'items' of information held by an individual. Information is, after all, not understood acontextually – it is held by people who interpret and integrate items of information within their own conceptual framework. Information is, therefore, effectively a complementary good, in that one piece of information may be needed to make sense of another.

If information is understood as existing only within individuals and as being interpreted by them, then the advantage inherent in attempting to 'lock up' know-how 'contractually' (Drahos, in Macdonald and Nightingale 1999) through the mechanism of confidentiality may be questioned as being contrary to the spread of innovations and therefore to the long-term interests of the economy as a whole.

The need for monopoly rights to foster incentives supports the notion of information as property and can also be seriously questioned as an economic proposition. Suggestions that a monopolist will always have greater incentive to invent than an open competitor have been doubted: Demsetz (1969) shows that this will only be the case where a linear model of two industries with equal output size prevails. Arrow (1962) even suggested that the inventor's incentive under competition will always exceed the monopolist's incentive. His analysis is particularly appropriate for confidences, as his model assumes that only the monopolist itself can exploit the invention. In any case, Yamey (1970) has shown that under more general assumptions, the competitor will have an incentive at least equal to the monopolist.

It is also misleading to suggest that an inventor can only receive value by restricting the use of his ideas. A source of the benefit from information based not in secrecy but in dissemination of information is noted by Hirschliefer (1971: 572). A firm can substantially profit from discovery of a new technology by acquisition of assets whose price will rise because of the adoption of the technology, followed by open and widespread disclosure. Thus he suggests that Eli Whitney could have profited from the cotton gin by buying cotton-producing land.

A particular cost of protection of commercial information is the inherent conflict with the aims of the patent system. One of the theoretical justifications for the patent monopoly is the encouragement to disclose the patented invention, which may then be used for experimental and non-commercial purposes as a stimulus to further invention (*Frearson v. Loe* [1878] 9 ChD 48 at 66 per Jessel MR). The original principle underlying grant of patent rights in English law has been the dissemination of information. For example, rights were granted conditional on the instruction of the people in an invention or process, 'because at first the people of the Kingdom are ignorant and have not the knowledge or skill to use it' (*The Clothworkers of Ipswich* case [1615] Godbolt 252, 78 ER 147). Newman in the Australian House of Representatives went so far as to state that the dissemination of information through the patent system was the very core of the whole innovative process (Hansard 1 May 1980). Yet to give information, initially confidential, the status of property without the need to publish would lock up information and undermine the patent system. On a random survey of the chemical industry in Britain, 22.2 per cent kept inventions secret which they would have patented if the law had not protected confidence (Kitch 1977). Hence secrecy fetters interfirm signalling, encouraging wasteful duplication of investment in innovation (Hirschliefer 1973). Under a regime of trade secrecy the competitive firm might never learn of a competitor's processes and would not learn of the technology incorporated into a new product until it was marketed. If the law is too zealous in enforcing secrecy, it will endanger some of the very goals that promoted the grant of protection – economic growth and productivity.

Adopting an information economic perspective leads to an ability to question the very nature of economic crime. This is particularly so where the basis of such crime lies in circumventing a legally imposed monopoly privilege (artificially imposed scarcity) where the economic rationale for that privilege may be questioned.

Prevention and Control Strategies

This has been only a shallow excursion into some laws and evidentiary issues that will arise in respect of industrial espionage, particularly in a telecommunications context. The remainder of this chapter is devoted to some ideas that may assist the generation of procedures to either help or thwart industrial espionage or to ensure that if it has occurred, an organization can get suitable recompense for the loss of the information. Securing information can be thought of almost in the same vein as securing one's home and contents against theft.

Compile an inventory of private information

According to a report by the Futures Group, 82 per cent of US companies with $10 billion or more in annual revenues have a formal competitive intelligence network (*Wired*, June 1988: 87). In order to effectively assess and protect information at risk, an 'inventory' must be taken of the organization's valuable information – itself a potentially risky process. Further to this, consideration must be given to those people and organizations who are, or could be, privy to the sensitive information. Once an assessment of the two preceding points has been made, the organization will have a database from which it can develop practices and procedures that can minimize the opportunity for industrial espionage in relation to their particular organizational requirements and structure. An objective of the practices and procedures ought to be to ensure that an audit trail is generated in respect of the transmission of information. The difficulty with such an audit trail is to ensure that it is composed of elements that would be admissible and persuasive as evidence in court. A real-life example of this can be found in a patent challenge case: in this case an employee had left a firm and six months later filed several patents which appeared to be for work he had been doing while employed by the firm. Effectively, he had taken their patent. The problem with the patent challenge provisions is that the challenger does not get to see the content of the patent (there is a chance that evidence could be manufactured ex ante to suit the action), so the challenge is 'blind'. Evidence that would have been useful here would be regular progress reports satisfying a particular template. The firm, however, had no such procedure in place. Employees may hate filling in forms, but surely firms dislike having millions of dollars worth of patent effectively stolen by ex-employees. In developing these practices and procedures it is important to contemplate the effect of their adoption on the organization as a whole. An important caveat is that some practices may be effective in preventing the misappropriation of the information, but they may also have adverse effects on productivity because of the imposition of transaction costs or damage to organizational morale.

The best-fit solution is one generated specifically for that organization. It must be reviewed periodically.

Individual organizations are best suited to identifying their private information. As a guide, however, valuable private information tends to be of the nature of (and this is not an exhaustive list): lists of suppliers and clients; lists of ingredients or chemical compounds; material not yet patented, or copyrighted material not made publicly available. Clearly, private information would also include databases; accounting data and accounts; any matter that could affect share prices or credit arrangements; and indeed anything that could affect goodwill (this might change rapidly) such as memoranda, letters, e-mails or their drafts.

Many of these materials are not what organizations might typically consider to be valuable, but the important consideration is the effect of the disclosure of the material on the organization.

Disclosure of such materials could result in damage such as: a loss of competitive advantage or the generation of new competitors; the loss of a statutory monopoly or damage to the goodwill of the organization. The consequence of such disclosure may damage an organization's share price or credit arrangements and result in evidence that must be delivered up to the opposing party in a legal action.

Restrict physical and logical access to private information

Generally speaking, an organization should make accessing private information without authorization particularly difficult, on the grounds that prohibiting such access is practically impossible. Access to buildings, lifts, rooms and systems is typically managed by a combination of magnetic stripe card (or smartcard) and PIN. If private information is able to be located in a secure room with access limited to a small number of people, misappropriation is less likely to occur. Ideally, such information should be stored in a location in such a fashion that a log is created to audit the comings and goings of individuals. Other considerations are video surveillance of the room. A low-tech answer is to ensure that the users have to pass a guard (or an administrator) to enter the room.

If storing the information in a central location is not possible, then other measures will provide an indication whether unauthorized activity may be occurring. Consider which staff require e-mail accounts and Internet access, telephone access, facsimile access and photocopier access rather than assuming that all staff require such services. Placing a PIN activation and counter on a photocopying machine will provide an indication whether an employee appears to be copying an inordinate amount of information. This is by no means infallible, but in conjunction with other measures it can prove effective. An organization's telephones can also be activated by a PIN as well as subject to restrictions for certain levels of call.

Reviewing telecommunications traffic (without accessing its content) can be a useful early warning system that might reveal undesirable information flows. Of course, this can act in a positive fashion – witness the birth of Java at Sun Microsystems. In that case the firm noted that several employees from seemingly disparate areas were exchanging a lot of internal e-mail. When questioned, the employees effectively stated that they were working on the creation of Java.

Confidentiality clauses

Confidentiality clauses will be difficult to employ after the fact of contracting; there must be further consideration in order to render such clauses effective at contract law. Despite the contractual situation, signatures on confidentiality clauses are good evidence for a breach of confidence hearing at equity.

With respect to contractually operative confidentiality clause damages, three options exist. The first is liquidated damages – here a set figure is arrived at and included in the clause. The danger with liquidated damages is that if the court cannot envisage the method by which the sum of damages was arrived at, it might hold the clause to be a penalty, thereby dismissing that sum. A similar form is by agreeing on the method by which damages will be calculated. Such a clause is best generated according to each specific element of secrecy, otherwise there is a risk of the clause being read as a penalty again. Relying on unliquidated damages may be risky also, as it will often require the court to consider expert evidence on how lost future profits could be calculated. The risk in some situations is that the court might consider the loss to be a mere loss of a chance.

Screening, selection and socialization

Often staff and others may have no conception that the information they deal with is private. Sometimes a solution to this will be to inform the staff about the sensitive nature of the information and ask them not to impart it to anybody, including family and friends. This will not always be the best approach, however, as occasionally it will act to make clear to staff that they possess something of value that could be exchanged. This type of educational strategy is really dependent on the nature of the work, the type of secret and the nature of the organization's culture.

In relation to new staff, the process of screening involves considering the type of individual that one's organization would like to attract and then taking time to frame the advertisement correctly. Consider the potential impact of personality in formulating selection criteria for positions. When examining resumes, actually ring former employers. Screening for employees who have goals and personalities suited to the organization is a useful way of helping to prevent any issue of industrial espionage. An overall aim should be to align the interests of the staff to the interests of the organization. Working environment and enjoyment are more likely to result in loyalty than mere monetary reward.

'Mark' information as private

Private information is also usefully protected, to a degree, by employing measures such as digital watermarks, digital fingerprints and encryption. Digital watermarking allows the owner of information to invisibly mark the information using a computer program. A digital fingerprint is a special kind of digital watermarking that is used as evidence. Encryption can ensure transmission security as well as storage security by scrambling the information so that it can only be read by a person applying the correct 'key', activated by a password. A weak point in such a system is the human element. Passwords are often lost or written on post-it notes, or are relatively simple to guess. The rule of an alphanumeric code with upper and lower case is often broken in favour of adopting the name of one's child or pet.

Encryption can be an especially effective measure when used in conjunction with digital watermarks. Digital fingerprints are a means of being able to track copiers. This makes it especially useful as a form of evidence of copying, as often a copier might not consider a need to alter the stolen image. Digital fingerprints and watermarking more generally, like encryption, are not infallible; several techniques exist to alter or remove such items. These techniques vary according to the way the software that creates the watermark is generated. Watermarking is also the term used for certain types of encryption that are currently the subject of numerous proposed standards in relation to digital music. In elaborate forms of watermarking, 'bots' (programs designed to hunt down the watermark) can be used to trace the path of the downloaded file. Such programs can, however, be deterred by good-quality firewalls.

Conclusion

The incidence of industrial espionage is no doubt likely to increase rapidly as a result of the proliferation of technological devices. This increase highlights the conceptual strain on legal systems seeking to legislate to protect pure data and information. The economic considerations in the protection of trade secrets and conflict with the objectives of patent are more telling where criminalization is requested. Overall, the best defence against industrial espionage is a well-prepared organization. No organization is invulnerable to misappropriation of information. Organizations should seek to make themselves less vulnerable targets by adopting sensible information and knowledge management systems. Organizations need to realize that information security should relate to the state of all of their data and information, not merely what is computerized. The information inventory and measures that are taken to protect information not only need to be re-evaluated periodically, but also to be constantly updated with each new development.

CHAPTER 10

The Electronic Misappropriation and Dissemination of Personal Information

Late in 1997, the director of the Australian Institute of Criminology was approached by an Australian government minister and asked to provide some briefing notes on the subject of sex slavery. The issue of illegal trafficking in persons, especially that involving an element of coercion, had become one of concern to the Minister's office. The director delegated the task to one of the authors of this book, who sought initially to see what might be learned on the subject from the Internet. Having entered the term 'sex slavery' on his favourite search engine, the intrepid researcher encountered a multitude of hits, most of which were entirely devoid of policy relevance. After further searching, he was able to compile a set of notes relating to the law of kidnapping and abduction in various jurisdictions, domestic and international, and forwarded them as requested. Relieved to be finished with the task at hand, he turned his attention to more mundane matters. But his research would return to haunt him: over the next month he received a number of unsolicited e-mail advertisements of a rather salacious nature.

It was obvious that the researcher left his electronic 'footprints' all over cyberspace, or at least in those corners of cyberspace wherein he had ventured. He had no intention of sharing his research agenda with anyone, least of all commercial entities interested in marketing pheromones and other products of little professional use to a criminologist. Nevertheless, his contact details were appropriated by a number of electronic entrepreneurs.

This anecdote is a metaphor for a larger issue: the erosion of privacy in the electronic age. Technologies of surveillance and tracking have given new meaning to the old adage 'You can run, but you can't hide'. Technologies of data-matching permit the construction of electronic dossiers on persons for a variety of purposes: some of them merely commercial, but others more sinister.

This chapter will review some of the motives and methods for the unauthorized appropriation of personal information. Clearly, personal information can take many forms, some of which have economic importance, while others have importance to our social and family lives. Using computers to acquire and disseminate personal information can, therefore, sometimes involve theft in the strictly criminal, property-law sense of the term. In other circumstances, the misuse of personal information acquired electronically might not readily be characterized as theft. For the purposes of the present discussion, however, we shall consider the unauthorized acquisition and dissemination of personal information as theft, even though it might not, strictly speaking, entail the commission of a property crime. Our discussion will also consider the interesting question of just what kind of personal details are in the public domain. It will also touch on the question of whether publicly available data can become transformed into private personal information. Finally, it examines how the electronic misappropriation of personal information should be regulated and in what circumstances it should be made illegal.

Personal information exists in many forms. Its misappropriation is by no means unique to the digital age. In 1361, the *Justices of the Peace Act* in England was directed at peeping toms and eavesdroppers. Of course the use of spies and informers long predated the emergence of English law, and with the rise of literacy, opportunities arose for the unauthorized acquisition of information from personal correspondence and diaries.

Perhaps the seminal work on individual privacy was the *Harvard Law Review* article published by Warren and Brandeis in 1890. One of their concerns was for the 'unauthorized circulation of portraits of private persons'. With uncanny prescience, they observed that 'Instantaneous photographs and newspaper enterprise have invaded the sacred precincts of private and domestic life; and numerous mechanical devices threaten to make good the prediction that "what is whispered in the closet shall be proclaimed from the housetops"' (1890: 195).

In this chapter we explore the appropriation of personal information using digital technology. As we shall see, the potential for electronic theft of personal information has never been greater. Developments in technology are facilitating this in some respects. Methods of so-called 'dataveillance' have been refined considerably since this somewhat ungainly term was coined by Roger Clarke in 1988 (see page 168). On the other hand, privacy-enhancing technologies have become publicly available. Powerful encryption techniques, often referred to as 'strong cryptography' and previously available only to governments, has now been democratized.

These trends have given rise to titanic struggles over law and policy, at least in Western industrial societies. Privacy advocates on the one hand contend that existing law offers the individual inadequate protection from government. By contrast, representatives of law enforcement argue that privacy-enhancing technologies are being exploited by criminals and terrorists, and that the legal tools available to government are inadequate to allow timely access to criminal information. Commercial interests, for their part, have mixed views. On the one hand they recognize that a modicum of privacy is essential to ensure public confidence in electronic commerce. On the other, they acknowledge the immense value of access to personal information for purposes of marketing. As our sex slavery researcher discovered, the commercial appropriation of personal data is becoming widespread in consumer societies.

It should be obvious to the reader that one could devote an entire book, or indeed a series of books, to the issue of privacy in the digital age (Diffie and Landau 1998; Rosen 2000b). After all, privacy merges into many other issues, and fields of inquiry, from ethics to economics to mathematics. Moreover, people's definitions of privacy differ, as do their own expectations of privacy. This chapter provides a selective discussion of issues relating to online privacy. Our concern lies mainly with appropriation of personal information rather than circumstances of intrusion, where one is involuntarily exposed to offensive content (see Grabosky and Smith 1998: Ch. 6). It is of course, not always possible to separate the two, as our sex slavery researcher was to learn.

In response to this perceived challenge, law enforcement and security agencies around the world are seeking and obtaining greater powers to intercept electronic communications. In democratic societies where the rule of law tends to prevail, these powers are generally subject to a degree of judicial oversight. In Great Britain, the *Regulation of Investigatory Powers Act* provides for monitoring and interception of electronic mail. In the United States, a system called 'carnivore' helps investigators to scan, capture and analyse e-mail packets (FBI 2000). In addition, a globally integrated communications intelligence system known as 'ECHELON' operated by the United States, Great Britain, Canada, Australia and New Zealand, permits the systematic collection and analysis of telecommunications content (Campbell 1999, 2000; Whitaker 1999; Associated Press 2000).

We largely exclude state surveillance and criminal investigation from the present discussion, though aspects of law enforcement are addressed in Chapter 9 on industrial espionage, and in Chapter 11.[1] Nor are we concerned with what might be called 'identity theft' (such as the theft of credit card details or social security numbers) for the purpose of committing subsequent fraud. These issues have been amply covered elsewhere (Smith 1999) in addition to their consideration in previous chapters. Theft of corporate or government information is discussed elsewhere in this book, as is theft of personal details in furtherance of extortion. Chapter 7 also highlights some of the ways in which database marketing may be abused. This nevertheless leaves a variety of circumstances in which personal details may be appropriated and misused, including marketing, embarrassment, intimidation, and voyeurism.

Technologies now permit the collation of otherwise disparate personal information in a manner which, while not actually constituting theft, allows the assemblage of considerable detail about an individual. Moreover, we define personal information rather broadly, to include photographic images and electronic correspondence – indeed, anything that may be digitized. And not only do these technologies allow the collection of details of personal characteristics, they also enable us to obtain a dynamic picture of a subject's movement, in real time or retrospectively. Technologies of tracking and location now permit the monitoring of persons and property. A variety of transactional information can be retrieved and analysed. The time and location of credit card transactions may now be recorded and archived, allowing the reconstruction of one's movements and spending patterns, while space geodetic methods of tracking through the use of satellites enable individuals to be located anywhere in the world as they make use of their computers. Where health records are maintained on databases, it is possible to ascertain an individual's entire medical history – clearly of interest to insurers.

The subject of privacy raises complex legal issues. To a greater extent than is the case with other issues discussed in this book, protection by means of the criminal law is limited. In some circumstances, civil remedies such as compensatory damages and injunctions to restrain objectionable conduct may provide effective relief, but in others an individual may have little recourse. Access to privacy-enhancing technologies can provide a modicum of protection for one's information, but the widespread availability of such technologies to members of the public may be seen as threatening to institutions of law enforcement and national security. Indeed, the protection of personal information is likely to remain one of the more intensely contested areas of public policy.

After a discussion of the motives for and methods by which personal information may be appropriated, we look at some of the remedies and countermeasures available to those who wish to protect their personal information. The chapter concludes with a discussion of whether and how a balance may be struck between the interests of the individual, the state, and the market. And again, as is the case with many of the forms of misappropriation discussed in this book, technical and market-based solutions will interact with legal remedies.

Motives for Personal Information Theft

Motives for the appropriation of another's personal information by one means or another are vastly diverse. Let us review just a few of these with some illustrative examples.

Social control

Violation of privacy has been a hallmark of authoritarian regimes. Surveillance fosters inhibition, and George Orwell's phrase 'Big Brother is watching you' has become a metaphor for surveillance as an instrument of social control.[2] As Foucault reminds us, knowledge is power. Surveillance yields information, which may be transformed into knowledge. It is now possible to determine the precise location of a mobile telephone (and the person using it, though not necessarily who that person might be). It is also possible to reconstruct the movements of that instrument (and presumably the person possessing it) over time. What we are often unable to do, however, is to know which human actor is making use of the electronic device being tracked. Technologies of biometric authentication may, however, make individual identification of the users of digital devices more effective than at present.

Of course, the state has no monopoly over social control. Employers, parents, spouses, and personal enemies often take a great interest in one's activities for the purpose of discouraging undesirable conduct. Surveillance is also one of the most significant instruments of social control in the contemporary workplace. Here, however, its objectives need not be sinister, and may include employee health and safety, fraud control, and the prevention of other undesirable conduct which may entail legal liability for the employer. Technologies of workplace surveillance range from video cameras to extensive monitoring of the use of information technology. In some locations, such as gambling casinos, surveillance, whether by direct observation, video camera or the systematic auditing of the turnover of each table and shift, is routine. This surveillance is unobtrusive but ubiquitous, and it comes with the territory – every employee knows it's there.

Of course the employees of a casino are not the only ones subject to surveillance. Patrons too are accorded scrutiny, for some of them may be tempted to engage in unfair play. Casinos may banish and bar players found cheating. Identification of those who may try to sneak back in on some later occasion will require scrutiny of all patrons as they enter the premises. Casinos, as other service industries, wish to appear customer-friendly, and therefore seek to make their surveillance as unobtrusive as possible. New technologies can enhance the unobtrusiveness of surveillance. Automated computerized face recognition now permits digital matching of a photograph taken on the spot with one that exists in a photographic library. This gives a further edge of meaning to 'Big Brother is watching you'.

There are, of course, other, more insidious forms of workplace surveillance. Every keystroke may be monitored. Every e-mail message sent or received and every website visited may be subject to review, as we saw in Chapter 4, where one enterprising woman maintained a database listing every Australian government user of her sexually explicit website. According to one commentator, 20 per cent of businesses in the United States report that they check employee e-mail randomly (Bennahum 1999: 102). The destination and duration of every telephone call may be logged. Obviously, employers have an understandable interest in knowing if their facilities are being used to peruse pornographic websites, for extensive personal communications, or for any other activity that may attract civil or criminal liability.[3] The basic issue here is the understanding reached between employer and employee at the time of engagement or on some later occasion.

Obviously, our sex slavery researcher had a legitimate reason for visiting some of the darker corners of cyberspace; for other individuals with other motives, research is able to provide a good cover for many clandestine activities.

Litigation

Many people who use electronic mail do so with unusual candour. In the words of Bennahum (1999: 102), 'E-mail is a truth serum'. But unlike a face-to-face conversation, electronic communications are not evanescent. Records persist, and may return to haunt one or more participants in the communication. Even when a message is erased, it may have been retained by another party to the communication, or it may have been 'backed up' on one or more system files, as Oliver North discovered when he was confronted with details of his involvement in the Iran-Contra affair (Walsh 1993). Indeed, technologies now permit the recovery from a hard disk of almost all information that has been erased.

However they may have been obtained, records of electronic communication can be useful in the resolution of disputes. One of the more vivid examples is the internal e-mail of Microsoft, which was adduced as evidence by the US government in its anti-trust action against the company (Rohm 1998). The government alleged that the company had engaged in predatory, anti-competitive behaviour, and subpoenaed company e-mail files to illustrate motives. One message read, 'If you're going to kill someone ... you just pull the trigger. Angry discussions beforehand are a waste of time. We need to smile with [the competitor] while we pull the trigger' (quoted in Rohm 1998). Rosen (2000a) reminds us that e-mail and other stored data may be 'bought, subpoenaed, or traded by employers, insurance companies, ex spouses and others'.

With seventeen million subscribers by mid-1999, America OnLine has received numerous civil subpoenas requesting subscriber information, often for divorce cases. But many communications are accessible merely by using readily available search technology. One litigant searched the Internet for his ex-wife's account name and collected 30 pages of messages that she had posted in chat rooms. Seeking greater visiting rights with his children, he presented them unsuccessfully to the mediator in his custody hearing (Glod 1999).

Commercial exploitation

There are many ways in which personal information may be exploited for commercial purposes. Excluded from consideration here are theft of credit card details, which are covered in Grabosky and Smith (1998: Chs 4 and 8), or theft of telecommunications services generally, discussed in Chapter 5 of this volume.

Whatever else its usefulness, personal information is undoubtedly of very great commercial value. In the United States, a great deal of personal information may be purchased from credit reporting companies (Gindin 1997: 1157). Nowhere is this more apparent than in marketing. Targeting prospective consumers becomes much easier in the digital age. Voluminous details abound, and are amenable to merging, matching and analysis as never before. Under some circumstances, this can be of great help to the consumer. Imagine receiving an e-mail reminding you that your wedding anniversary is the next day, and inquiring whether you would like flowers delivered to your spouse on the occasion. Information retrieval technologies also facilitate the location of long-lost friends and may facilitate family reunions.

But electronically informed marketing does have its downside. There are those who find commercial information, delivered by whatever medium, to be intrusive. Then there are people who may be embarrassed, or worse, when the commercial information in question comes to the attention of a third party. We recall the anecdote (perhaps apocryphal) about the telecommunications carrier which offered special discounts to customers who regularly call certain numbers. On one occasion, a housewife received a regular telephone bill with such an offer enclosed. In this case, the specified number was unfamiliar to her. She rang it up only to find that the subscriber was having an affair with her husband.

In addition, there are those websites that are directed at children and that collect personal information about children and their families, often in conjunction with contests or other activities relating to the site. Some parents would object to their children's exposure to Internet advertising.

In addition to marketing, personal information can inform commercial risk management. Our credit history is of considerable value to a subsequent provider of credit. Our health history can be of great interest to an insurer or employer. All else equal, a person who is HIV-positive or who has a history of depressive illness may be a less attractive employment prospect. Even for the person in perfect health, her employment history, or indeed even the chat rooms she frequents in cyberspace, can be of interest to a prospective employer.[4]

Personal data are thus becoming a valuable commodity. Persons with access to databases have exploited this for their own financial benefit – sometimes legally, sometimes not. The NSW Independent Commission Against Corruption identified many instances in which individuals in the public service sold nominally confidential personal data for their own advantage (ICAC 1992). Spink (1997) identifies a number of cases in which police in the United Kingdom engaged in the unauthorized disclosure of information for the benefit of commercial interests.

Some Internet service providers have been known to buy or sell subscribers' personal information. In 1999, the Attorney-General of Minnesota alleged that a bank sold customers' private financial details to a telemarketing company for US$4 million plus commissions. The details were extensive and were reported to have included telephone and social security numbers, marital status, homeownership, bankruptcy status, credit card details, bank account balance and information on recent transactions (*Corporate Crime Reporter* 14 June 1999: 14, 24).

Online commerce greatly facilitates the development of customer profiles for marketing purposes, especially when the user provides additional personal information as through filling out a questionnaire. It is now possible for an online advertiser to identify a visitor to a given website, then sift through their databases and serve targeted ads on the visitor instantaneously.

Online marketers are also collating data obtained from more traditional mail order lists, and in some cases these marketers develop strategic partnerships. In June 1999 the Internet advertising firm DoubleClick announced plans to merge with Abacus Direct, a company that sells information about consumers' catalogue purchases (Tedeschi 1999). Using 'cookie' technology (discussed below), a website can track visitors, identify them, and merge their details with a vast amount of personal information accessed from other databases. The joint venture formed in 1999 between Australia's Publishing and Broadcasting Limited and US-based Acxiom Corporation, is further indicative of the marketing potential of these 'customer relationship management technologies', as we saw in Chapter 7.

That personal information has value is not only implicit in the fact that many commercial institutions actively collect it. Increasingly, companies are willing to pay individual customers or subscribers, in cash or in services, for personal information. And those who can obtain the information without paying for it will do so.

Voyeurism

Strictly speaking, the word 'voyeurism' refers to a person who derives gratification from observing another person in a state of undress, or engaged in sexual activity. The practice is hardly new; as we have observed, peeping toms have been around a long time. It is but a small step from direct observation to indirect observation by means of photography. While early photographic technology was bulky and obtrusive, making clandestine photography rather difficult, advances in miniaturization have greatly facilitated covert photographic activity. More recently, the advent of digital technology permits the timely production and widespread dissemination of such imagery. One need only do a Web search on the terms 'voyeur' or 'upskirt' to grasp the point.[5] Ostensibly candid photographs taken with hidden cameras in public lavatories are now accessible to anyone with a computer and a modem; the use of concealed cameras by male patrons of shopping malls in the United States has attracted police attention (Davis 1998).

Prominent people often go to great lengths to avoid photographers, whether simply because they wish to be left alone or because they are concerned that embarrassing photos will be published. Were they alive today,

Jacqueline Kennedy Onassis and Princess Diana could describe not only the harassment they experienced at the hands of paparazzi but also the discomfort of seeing their unflattering image on the front page of a tabloid newspaper.

Voyeurism need not be limited to visual images. We use the term more broadly here to refer to collection of personal information in whatever form for the purpose of titillation. One need only think of the public fascination with the private conversations of Princess Diana as recorded on the so-called 'Squidgy' tapes; private correspondence in whatever medium may also be of interest to the curious. And now the dissemination of such data is easier than ever.

The collection of personal information for its own sake may entail more than simple curiosity. Bernstein (1996) recounts a case involving inmates of a correctional facility at Lino Lakes, Minnesota, who compiled an extensive database on children from the surrounding area. The prisoners, who had access to information technology through a prison-based computer programming and telemarketing business, scanned children's photographs and collated other information from local newspapers. The annotated files on local children contained information about which girls took piano lessons and which ones had entered children's beauty contests, and also included descriptions of children's physique. In addition there were annotations on some files referring to 'latchkey kids', 'cute', 'Little Miss pageant winner', and the existence of 'speech difficulties'. The towns in which the children lived were alphabetized and coded with map coordinates.

Whether these data were collected purely for purposes of voyeurism or fantasy, or for the planning of subsequent criminal activity following release, or were compiled for sale to child molesters, one can see how ostensibly innocent public information may be gathered, collated, and subjected to misuse.

Political protest or intimidation

Some individuals will publish another's personal information as a political statement, or arguably in a manner that may be perceived as veiled intimidation. In this latter sense, the distinction between protest and social control is blurred. One Australian firearms enthusiast, angered by the introduction of national firearms licensing and registration requirements in Australia, and by bipartisan support for the prohibition of most semi-automatic firearms, sponsors a 'Government Home and Garden Competition'. In addition to containing strident anti-government messages, the website offers a prize for 'the best home photograph and address of your favourite politician or public servant' (Lock, Stock and Barrel 2000).

A resident of Washington State created a website containing personal information about public and private figures that he feels have wronged him (Barry 1997). Those mentioned include judges, lawyers, politicians, and employees of a collection agency. Personal information disclosed in the site includes home phone numbers, maps to individuals' homes, and photographs, including photos of people's cars. The impact is heightened by a small cartoon character who struts across the screen and appears to urinate on the photograph (Sheehan 1999).

Others have posted real-time images of the home of then British Defence Secretary Michael Portillo to protest the British Government's contention that there can be no expectation of privacy in public places (Davies 1997: 155, n. 47). On 26 November 1995, in order to emphasize the insecurity of online medical data banks, the *Sunday Times* revealed how confidential medical records of prominent people in England had been illegally obtained from National Health Service staff and doctors' surgeries for £150 (Smith 1998: 194).

Personal information may be sought and used to discredit an individual, for political or other reasons. The controversial nomination of Judge Robert Bork to the Supreme Court of the United States prompted one inquisitive soul to obtain the record of Bork's transactions at a local video shop. While the use to which such data were to be put was not immediately evident, one doubts that it was for the purpose of writing what journalists would refer to as a human interest story. As it turned out, the judge's viewing preferences were rather conventional.

Stalking

The collection of personal information (whether by digital or other means) may eventually lead to an attempt to make contact with the person in question. The contact may be direct and personal, or more anonymous. In some cases, it may be fatal. In 1989 the actress Rebecca Schaeffer was murdered by a stalker who obtained her residential address through the California Department of Motor Vehicles.

The pursuit or threatening of a victim may also entail digital technology. In 1998, about 20 per cent of the 600 cases reviewed by the Los Angeles District Attorney's Stalking and Threat Assessment Team involved some form of e-mail or electronic communication (Miller and Maharaj 1999). There are a number of reasons why digital stalking may be expected to increase. First, the continuing uptake of digital technology means that a greater number of potential victims and offenders will have access to the relevant technology. In cyberspace, as elsewhere, crime follows opportunity. Then, as noted above, digital technology affords the illusion if not the reality of anonymity. And finally, the abundance of information available about ordinary individuals, where they go, what they do, and with whom they associate, seems destined to increase.

Harassment

Personal information may also be appropriated for purposes of harassing the subject. One man allegedly stole nude photographs of his former girlfriend and her new boyfriend and posted them on the Internet, along with her name, address and telephone number. The unfortunate couple, residents of Kenosha, Wisconsin, received phone calls and e-mails from strangers as far away as Denmark who said they had seen the photos on the Internet. Investigations also revealed that the suspect was maintaining records about the woman's movements and compiling information about her family (Spice and Sink 1999).

In another case a rejected suitor posted invitations on the Internet under the name of a 28-year-old woman, the would-be object of his affections, which said she had fantasies of rape and gang rape. He then communicated via e-mail with men who replied to the solicitations and gave out personal information about the woman, including her address, phone number, details of her physical appearance and how to bypass her home security system. Strange men turned up at her home on six different occasions and she received many obscene phone calls. While the woman was not physically assaulted, she would not answer the phone, was afraid to leave her home, and lost her job (Miller 1999; Miller and Maharaj 1999).

One former university student in California used e-mail to harass five female students in 1998. He bought information on the Internet about the women using a professor's credit card and then sent 100 messages including death threats, graphic sexual descriptions and references to their daily activities. He apparently made the threats in response to perceived teasing about his appearance (Associated Press 1999c).

Mistake

It is often said in the aftermath of a mishap that, when faced with a choice between an explanation based on conspiracy and one based on a mistake, it is more likely than not the latter which is most valid. Personal information is often disclosed inadvertently. The recipients may choose to ignore it, or may appropriate it for one or more of the motives discussed above.

In April 1999, Yahoo inadvertently revealed the addresses and order information of customers of Vitanet, a nutritional supplement vendor that operates through the Yahoo site. The exposed information included partial credit card numbers, products ordered, amounts spent and a link to a map. The map link included customers' street addresses and a map of their surrounding area. The disclosure resulted from a software bug, which was immediately repaired (Wolverton 1999a).

AT&T accidentally disclosed the e-mail addresses of 1800 customers who received an e-mail about their new rates on international calls by keeping all recipients' addresses in the 'to' field. Each person who received the message also received the e-mail addresses of all the other customers who received the message. A similar incident occurred at Nissan, when 24000 e-mail addresses were posted with an e-mail advertising the Nissan Xterra, a new sport utility vehicle (Wolverton 1999b, 1999c).

Of course, the private sector has no monopoly on incompetence. In February 1999, police in Britain inadvertently posted on the Web the names and addresses of people who had come forward with information about a notorious racially motivated murder. The accused themselves were among the recipients of the report. The informants were offered police protection (Reuters 1999).

Data on thousands of patients of the University of Michigan health system were inadvertently made available on the Internet for a period of at least two months, until a student discovered the error by chance in February 1999 (Upton 1999). The lapse occurred when the hospital was developing a new patient-scheduling system. The company installing the system was given access to what was thought to be a secure server, but the data were discovered by a student seeking to locate a physician. In another case in Pinellas County, Florida, the names of almost 4000 people suffering from AIDS were disclosed when a computer database was inadvertently made public (Smith 1998: 194).

On other occasions, ISPs have erred in disclosing subscriber details to unauthorized parties. Perhaps the most widely publicized case to date has been that of US Navy Chief Petty Officer Timothy McVeigh (unrelated to the Oklahoma City Bomber). Without lawful authority, Navy investigators sought to obtain the identity of an America OnLine subscriber who was using the online name BOYSRCH, and who had indicated his sexual preference as gay. In the absence of a subpoena, an AOL employee identified the subscriber as McVeigh, who was subsequently disciplined for violating the Navy's 'don't ask, don't tell' policy. McVeigh successfully challenged the Navy's action and was allowed to retire with an honourable discharge (*McVeigh* v *Cohen*, 983 F. Supp. 215 DDC, 26 January 1998).

But the capacity for inappropriate disclosure of personal information is significant. Kang (1998: 1220) relates how an investigative journalist, using the name of a notorious convicted child murderer, obtained the addresses and telephone numbers of 5000 children from a database marketing firm.

The above categories may not be mutually exclusive. Mistakes can lead to voyeuristic invasions of privacy, or to commercial exploitation. CNN (1999) reported that following a security flaw in Microsoft's Hotmail service, a Swedish business manager's communications with two prostitutes were published on an anonymous website. The names and telephone numbers of several of the prostitutes' clients were included. In Japan, during the course of a software conversion, the information systems of a major bank were penetrated, and personal details of up to 20 000 customers were stolen. A number of files were then provided to a mailing list vendor in Tokyo (Reuters 1998).

Technologies of Information Theft

Roger Clarke (1988a) uses the term 'dataveillance' to refer to 'the systematic use of personal data systems in the investigation or monitoring of the actions or communications of one or more persons'. In the past, information privacy was protected by data dispersion. A great deal of personal information may have been stored here and there in various locations (whether public or private), but aside from major investigations, the cumbersome logistics of sorting through rooms full of forms in one place and another precluded collation on any significant scale. Technologies of data manipulation that permit merging of databases and matching of individual identities now facilitate the aggregation of data from disparate sources (Clarke 1998a). The term 'data mining' is commonly used to refer to such practices. The linking of disparate data is facilitated by the existence of identification numbers, which are common in most industrial societies; the nine-digit Social Security Number in the United States is a classic example. Through the collation of disparate personal details, the whole becomes greater than the sum of its parts.

By way of modest illustration, one can imagine the difficulty 50 years ago of finding a residential address in a printed telephone book when all one had was the telephone number. One is now able to interrogate the 'electronic white pages' by number, or interrogate numerous telephone databases by name (see also Gindin 1997: 1156). Indeed, one entrepreneurial investigator in the United States advertises commercial services that would enable one to obtain: the name and physical location of another telephone subscriber if one provides a telephone number;[6] another's telephone number if one provides a physical address;[7] the dates and interstate long-distance numbers called by another subscriber if one provides that subscriber's name address and telephone number; and another's credit card transactions if one provides that individual's credit card number, name, address, telephone number and social security number. The website advises that it supports anti-stalking laws, and that the request must be for a legitimate purpose (unlistedphonenumbers.com 2000).

Today's technologies permit scrutiny of e-mail, at least to the extent that it is archived by organizations or service providers. *Deja.com* indexes all usenet postings and facilitates a search of archived messages. Lessig (1995: 1748) describes how UNIX-based operating systems allow ordinary users to perform a wide range of monitoring functions.

The information technology industry has developed a variety of means to facilitate the capture of personal information. Intel, the world's largest chip-maker, released a product in January 1999 that included a serial number enabling the identification of users and their location. The ostensible justification for including the feature in the new Pentium III chip was the verification of user identity, and thus the enhancement of security in electronic commerce. A threatened boycott of Intel's products led the company to provide a facility for the user to disable the feature (San Jose Mercury News 1999).

Microsoft, as part of its Office 97 software package, provided for the incorporation of a unique identifier number and other details in the properties information of Office 97 documents.[8] In addition, its registration procedures allow customers to send hardware configuration information. This can expedite the resolution of a product support call. The 'Registration Wizard', as it is called, allows customers to review all information sent to Microsoft, as well as providing the option not to send hardware configuration information. Microsoft advises that 'it does not use this hardware information for any marketing or user tracking purposes'.[9] Elsewhere, the term 'mouse droppings' has been applied to the traces left by a visitor as she navigates cyberspace with the aid of a mouse (Byford 1998: 48).

Another technology is colloquially called 'cookies' after uneaten biscuits left in the jar (Clarke 1998c; Kang 1998). A feature of Netscape's web technology, a cookie is information recorded and stored on a user's computer. When Alice, through her Web browser, gains access to a Web page, the server that hosts the page sends back the page Alice requested, *plus* an instruction to write information in Alice's computer. Whenever Alice returns to the requested web page, the information in her cookie is transmitted to the host server. Alice may find this advantageous, as it may facilitate her interface with the requested page by indicating her interests. On the other hand, the existence of cookies permits interrogation of one's computer regarding sites visited. Alice may not wish to share knowledge of her visiting habits with anyone, in which case she will have to take positive steps to disable her cookies.

Remedies

The 'numerous mechanical devices' heralded over a century ago in the seminal work of Brandeis and Warren have become a fact of life, and the protection of personal information in the digital age is a daunting challenge. Both governments and large commercial entities value personal information. At the same time, it is impossible for a citizen to function normally, and it is becoming increasingly difficult for one to exist at all in advanced industrial societies without leaving electronic footprints. Institutions and instruments of privacy protection are numerous and diverse, but hardly comprehensive. We discuss some of the basic means of protecting privacy in ascending order of legal coerciveness. No one of them can be regarded as a panacea, but in combination they begin to provide a foundation for privacy protection (Raab 1997).

General public awareness

Basic knowledge of privacy and threats to privacy are the foundation on which all of the institutions discussed below are based. The work of Westin (1967) helped place the issue of privacy on the policy agenda in Western democracies at the beginning of the information age. The role of citizens' groups such as the Electronic Privacy Information Center in the United

States is illustrative. The Australian Parliament defeated legislation in 1988 that would have introduced a national identity card. In a 1995 Harris Poll, 80 per cent of Americans reported that they had lost control over personal information (Davies 1997: 147).

Governments themselves may assist members of the public to become aware of privacy issues by appointing commissions of inquiry (Jackson 1997: 160–1; Great Britain 1990), establishing specialized privacy 'watchdogs' or publishing relevant information through consumer affairs agencies. The first government inquiry into general privacy issues in Australia was conducted for the State of New South Wales in the early 1970s (New South Wales 1973). New South Wales, the largest state in the Australian federation, has established a Privacy Commissioner (New South Wales, Attorney-General's Department 2000) and the Australian Commonwealth has a Federal Privacy Commissioner (Australia, Privacy Commissioner 2000).

Self-help

Individuals are not totally powerless to protect their privacy in these modern times. Common sense dictates basic practices. Just as one would be ill-advised to speak in full voice in a roomful of strangers, one should be cautious about baring one's soul electronically. Responsible parents will supervise their children's use of information technology, and will caution them about divulging personal details to commercial interests or to strangers. Websites abound with information for parents to assist in their children's safe navigation of cyberspace.

Technology is a double-edged sword. Just as it provides unprecedented ways of violating privacy, so too does it enable individuals to defend their own privacy. So called 'Privacy Enhancing Technologies', once the monopoly of governments, are now widely accessible to ordinary individuals (Denning 1999). 'Strong' cryptography is becoming a standard feature of many IT products. Technologies of anonymity and pseudonymity enable users to conceal their identity from all but the most determined and best resourced investigators.

Ways are also becoming available of neutralizing, at least to a degree, some of the technologies of dataveillance discussed above. Technology can now facilitate the involvement of individuals in decision-making about the disposition of personal information. Microsoft is developing software that will allow consumers some control over the personal details they disclose in the course of electronic transactions. The newer search engines, for example, notify users before cookies are set. They may also incorporate 'cookie disablers' which 'switch off' the application. Various technological means are also available to protect computer hardware from electromagnetic radiation surveillance, though these tend to be beyond the means of all but the most well-resourced individuals and organizations. Rosen (2000a) describes one of his former students who has special personal firewalls installed to detect whether anyone is spying on him with a hacker program such as BackOrifice and NetBus, as well as a security package called Kremlin, which encrypts personal documents and obscures any partially deleted files on his hard drive by masking them with zeroes and ones.

Industry self-regulation

In the domain of electronic commerce, as in others, the spectre of government intervention has moved industry to introduce a variety of initiatives to regulate its own affairs. Increasingly, large Internet sites now tell visitors how their personal information will be used (America OnLine 2000). Relatively few sites, however, give users the chance to opt out of having their information collected or allow them to review their personal details. Nor do all sites promise to keep the information confidential. The Platform for Privacy Preferences (1999) was established by the World Wide Web Consortium Project to develop a model for disclosure of data collection and disclosure practices by websites.

But self-regulatory initiatives are no guarantee of compliance. One Internet company, Geocities, offered its clients free space for web pages in return for personal information. Geocities had promised that this information would not be distributed without the customer's permission, but had given this information to advertisers without customer consent. The case was brought before the US Federal Trade Commission and was settled with Geocities agreeing to several conditions. These were to rewrite their privacy statement, to obtain parental permission before collecting information from anyone under 12 years of age, and to provide a clear link to the FTC's website that contains information on privacy rights (Associated Press 1998b).

Commercial influences

These measures are in turn reinforced by pressures and signals emanating from the market. Consumers may be in a position to identify and resist technologies and practices that threaten personal privacy. The revelation of certain commercial products or practices that might threaten personal privacy usually generates a consumer backlash, reinforced by organized privacy advocates, whose capacity to inform and to mobilize the public should not be underestimated. We have already noted the deference of Microsoft and Intel to privacy concerns arising from the incorporation of identification technology in new products. So it was that AOL cancelled plans to make subscriber information available to its partners for telemarketing. In response to privacy concerns, some service providers have removed sensitive information from their databases. In 1996 Yahoo removed a reverse telephone number search facility from its service. Lotus abandoned a project to market a database of personal information on over 100 000 000 residents of the United States (Gindin 1997: 1160, 1184).

In some settings, market forces may contribute to the protection of consumers' personal privacy. IBM, second only to Microsoft as an advertiser on the Web, announced early in 1999 that they would withdraw advertising from sites that did not have clear policies on privacy (Girard 1999). The requirement took effect on 1 June. Within three weeks, the company said that 98 per cent of the websites where it advertises now carry a privacy statement. Not long thereafter, Microsoft followed suit, going one step further by proposing to check the details of privacy statements on the more than 750 Internet sites (as of June 1999) where it places advertisements. Microsoft will in particular assess the site's compliance with guidelines set down by the FTC: advising visitors that information is being collected; asking consumers for consent; giving them access to those details; and maintaining security of the data (Associated Press 1999c).

Commercial influences may also be enhanced by the services of third parties that specialize in the independent certification of compliance. Shapiro (1987: 205) refers generally to 'private social control entrepreneurs for hire'. And so it is that a market has begun to emerge for privacy certification services (Dancer 1999) in much the same way as we saw in Chapter 7 with the certification of business sites. One company, TRUSTe (1999), assists Web publishers to develop privacy policies and offers compliance certification for a fee. The client may then display the 'Trustmark' on its website. TRUSTe also conducts audits of their clients and invites consumer complaints, which may then lead to investigations. The service also provides additional guidelines for sites directed at children under the age of 13, such as an undertaking to get parental consent to obtain information. Compliance with these special requirements enables the site to display a special seal. It might be noted that TRUSTe is not universally embraced by privacy advocates, who tend to favour strong legislative protections over self-regulatory initiatives.

Common-law remedies

In some jurisdictions, common-law remedies may be available to those whose privacy has been violated. Causes of action may include invasion of privacy, trespass, and the appropriation of name or likeness when the use is not reasonably related to a matter of public interest. Injunctive remedies and the use of defamation proceedings also provide powerful weapons for those intent on protecting private information or who seek financial redress for electronic wrongdoing.

But common-law remedies may be limited, as they are in the United States and Australia (Hauch 1993; Jackson 1997).[10] Where they exist, they may be counterproductive. In one case, a woman who provided personal details for a marketing survey in return for free products and discounts received a sexually explicit and offensive letter. The survey had been processed by prison inmates, a number of whom did not feel bound by privacy conventions. The woman sued the marketing company that had fielded the survey on the grounds that it did not exercise appropriate supervision of its contract employees (Bernstein 1997). The company accessed its database to assist in the preparation of its defence and retrieved an abundance of information about the plaintiff, including 'not only her income, marital status, hobbies and ailments, but whether she had dentures, the brands of antacid tablets she had taken, how often she had used room deodorizers, sleeping aids and haemorrhoid remedies' (Bernstein 1997). Some of this information was used to test the veracity of the plaintiff in the course of a lengthy deposition. Ironically, privacy litigation can lead to further loss of privacy.

Legislation

Legislative protection for personal information is not new. As early as 1776 the Swedish Parliament enacted the *Access to Public Records Act*, which restricted the use of government-held information. In the modern era, the first data protection law was enacted in the German state of Hesse in 1970, followed by national laws in Sweden (1973), the United States (1974), Germany (1977), France (1978) and the United Kingdom (1984) (Flaherty 1989). At present over 40 jurisdictions around the world have such laws or are in the process of enacting them (Banisar 1999).

In Australia, the only state or territory to have a specific privacy law is New South Wales, which enacted the *Privacy Committee Act* 1975. The *Privacy Act* 1988 (Cwlth) lays down strict requirements which federal government agencies, and those of the Australian Capital Territory, must observe when collecting, storing, using and disclosing personal information, including that held electronically. The Act also gives individuals access and correction rights in relation to their own personal details.

In general, privacy initiatives tend to be limited in scope. The *Australian Privacy Act*, for example, applies only to agencies of the Federal and Australian Capital Territory governments. In the United States, the *Electronic Communications Privacy Act* of 1986 does not protect information that is readily accessible to the public or for which consent of one of the parties has been obtained (Byford 1998: 58). The Privacy Amendment (Private Sector) Bill 2000, introduced into the Australian Parliament on 12 April 2000, would create a national scheme of co-regulation to provide for the collection, holding, use, correction, disclosure and transfer of personal information by private sector organizations. The Bill provides a default framework for the protection of personal information, which will bind private sector organizations unless they have their own privacy code that has been approved by the Privacy Commissioner.

In most nations, legislative response to emerging threats to privacy may be characterized as piecemeal, and the metaphor of patchwork is not inapposite (Byford 1998: 57). Gindin (1997) reviews the mosaic of legislation in the United States, which includes, *inter alia*, separate provision for protecting the identity of intelligence agents, video rental records, and driver's licence details.[11] By contrast, medical records have yet to be accorded protection under United States law (Cate 1998). Some jurisdictions are solicitous of children's vulnerability to online violations of privacy, and require that online accounts must not be provided to children without the consent of a parent or responsible adult, who are encouraged to supervise and control children's access to Internet content (*Broadcasting Services Amendment [Online Services] Act* 1999 [Cwlth]). The United States requires all websites gathering information from children under 13 to get verifiable parental permission first, and to post clear policies about how that information is used (*Children's Online Privacy Protection Act*).

The emergence of new law usually follows the occurrence of a prominent unfortunate incident. The disclosure of Judge Bork's video rental records inspired the *Video Privacy Protection Act* (18 USC § 2710 et seq.). This Act provided for civil remedies, including punitive damages, for unauthorized disclosure of rental details. The *Driver's Privacy Protection Act* (18 USC Chapter 123), enacted in the aftermath of Rebecca Schaeffer's murder, bars US state governments and their employees from selling or releasing personal information such as social security numbers, photographs, addresses, telephone numbers and birthdays without the subject's consent.[12]

The international benchmark for information privacy has been set in Europe (see Mayer-Schonberger 1997). The 1981 OECD Guidelines on the Protection of Privacy and Transborder Flows of Personal Data inspired the 1995 EU Directive calling on member states to enact specific data protection measures. Significantly, they prohibit export of personal information outside the EU except to countries offering an adequate level of data protection (for a discussion of EU data in comparative perspective see Cate 1998). The extent to which these high European standards will prevail in the global economy remains to be demonstrated. Indeed, it might be argued that the EU Directive is a 'mainframe' solution to a problem which is to a very great extent 'distributed.' It misses the point that micro-data situated all over cyberspace can be accessed and collated with unprecedented ease.

In the United Kingdom, the *Data Protection Act* 1998 (Eng.) greatly enhances individuals' rights to information privacy. The Act gives a person the right to prevent processing of his personal details for purposes of direct marketing, or to prevent processing of information likely to cause damage or distress. Individuals are entitled to compensation from the data controller for damages resulting from a breach of provisions (Bainbridge and Pearce 1998). However, to ensure that law enforcement agencies continue to be able to get access to data under lawful warrant, the Regulation of Investigatory Powers Bill 2000 was introduced into the UK Parliament. The legislation seeks to ensure that technologies such as encryption cannot be used to defeat lawful surveillance and interception of communications (United Kingdom 2000).

Criminal sanctions

As we have seen, the criminalization of certain types of private surveillance dates to the fourteenth century. Some of the violations of privacy which authorities define as most egregious may attract criminal sanctions. In both the United States and Australia, for example, one can go to prison for revealing the identity of an intelligence agent.[13]

In some jurisdictions, laws have been enacted to prohibit certain methods of information theft. The law of Virginia now includes a provision that prohibits 'unlawful filming, videotaping or photographing of another'. The offence, which applies to circumstances in which victims had a reasonable expectation of privacy, carries a maximum penalty of twelve months in jail. In Wisconsin, the law prohibits taking photographs of a person nude without their consent.

Laws on stalking and solicitation of sexual assault are usually broad enough to prohibit the online manifestations of such behaviour, though one may expect further refinement of laws to criminalize, whether necessary or not, threats sent via pagers, e-mail, faxes, voicemail and other electronic communications.

Conclusion

The loss of one's private life is often accompanied by a decline in spontaneity, creativity, and a diminished sense of self (Byford 1998; Kang 1998). When Big Brother is watching, one is less able to manage one's identity; one loses power vis-à-vis the individual or organization that commands knowledge. At the beginning of the new millennium, it is interesting to note that the state is no longer the only major threat to individual privacy. Governments have been joined, if not surpassed, by large commercial organizations which would exploit, legally or otherwise, the increasing volume of personal details that accumulate in cyberspace.

As the accelerating pace of technological change exceeds that of public awareness, which in turn exceeds that of the law, privacy remains exceptionally vulnerable. Personal information may now be obtained from an unprecedented variety of sources in an unprecedented variety of ways, not all of which are illegal. The individual who covets, in the words of Brandeis and Warren, 'the right to be let alone' faces a dim future in the digital age. There is, at present, no reasonable expectation of privacy for messages posted on the Internet, because to a significant extent, 'on today's Internet, people *do* know you are a dog' (quoted in Gindin 1997: 1177).[14]

Personal privacy has become and is destined to remain one of the most strongly contested areas of public policy in democratic societies. It seems likely that government access to personal information will remain strictly circumscribed, at least in theory. While governments will continue at least to play lip service to the importance of privacy, they will maintain that a degree of access to personal information is essential for law enforcement and national security. Commercial access to personal information is another matter, however.

Governments remain reluctant to constrain the activities of business, particularly in industries such as information technology, which hold such promise for economic growth. At the same time, it is becoming increasingly apparent that uncontrolled access to personal information may inhibit the full exercise of individual political and economic freedom. Not only can dataveillance be bad for social life; it can also be bad for commerce. A modicum of privacy is essential for the foundation of trust on which electronic commerce must be based. Commercial interests have begun to recognize that unlimited access to personal details may have a chilling effect on Internet commerce. It thus appears that an optimal solution will be reached in which some information is available, but regulatory and quasi-regulatory institutions will limit untrammelled access.

It may be that the solution to the problem of commercial exploitation of personal information will be an economic one (Hagel and Rayport 1997). Many individuals will gladly reveal abundant personal information if they think there is something in it for themselves, whether a discount, preferential service, or a chance to win a grand prize. To the extent that this exchange is formalized, privacy will be transformed from a right to a commodity (Davies 1997). Then the interests of business and consumers may begin to converge, and there will be a business case for protecting consumer privacy. Whether the commodification of privacy will lead to the disclosure of even more personal information than at present is an interesting question.[15]

The law has yet to catch up with the technology of dataveillance and data-mining. To the extent that a person has some form of property right in personal information, its nonconsensual appropriation by another might properly be termed theft. In any event, there can be no single solution to the issue of privacy. If there is to be anything approaching privacy in the digital age, it is likely to entail a combination of legislative protection, self-help, and market solutions.

Notes

1 The illegal interception of telecommunications is discussed in Grabosky and Smith (1998: Ch. 2). Legally condoned searches, of course, can also be very intrusive. Rosen (2000b) describes the unsent love letters discovered in a search of Monica Lewinsky's computer by investigators from the Special Prosecutor's Office.

2 Systematic surveillance of political adversaries is a common practice. The FBI conducted very intense surveillance of the famous civil rights leader Martin Luther King Jr, including the use of listening devices in King's hotel rooms. The surveillance revealed a great deal about King's private life, which the FBI director, J. Edgar Hoover, sought to use to discourage King's continued political activity, or ultimately, to diminish his legitimacy (Garrow 1999).

By labelling individuals as belonging to a particular class, the state may seek to constrain their activities, or at least enable citizens to exercise surveillance on their own behalf. The State of Virginia has an online sex offender registry which allows one to search by name and location (county, municipality or postcode, and provides photograph, address, age, height, weight, hair colour, and lists the subjects' convictions. The website cautions that 'unlawful use of the information for purposes of intimidation or harassing another is prohibited and willful violation shall be punishable as a Class 1 misdemeanor'. The site contains information on over 4600 offenders; in the initial six months of operation, it reported having undertaken over five million searches on behalf of over 845 000 visitors.

http://sex-offender.vsp.state.va.us/Static/Search.htm (visited 27 November 2000).

3 This might include the sexual or racial harassment of other employees, the communication of defamatory information generally, or the disclosure of confidential corporate information. See Hart (1998).

4 There are those who contend that transparency of personal information is a positive good. See Schwartz (1997).

5 A search in November 2000 using the term 'upskirt' http://www.alltheweb.com/cgi-bin/search?type=all&query=upskirt (visited 27 November 2000) yielded 1 112 376 matches, the contents of which were not subjected to analysis.

6 A clever consumer of these services will soon discover that one can do this at no cost, by availing oneself of a facility called reverse lookup. See http://in-114.infospace.com/_1_7018763_info/reverse.htm (visited 18 June 1999).

7 This too, can be obtained *gratis* from the reverse lookup website. For an additional US$39, one may request a search for
Current and previous addresses going back ten years
Any additional phone numbers available
Driver's license physical descriptions in Florida and Texas
Family members of individual
Other people at the same address
Neighbors with listed phone numbers
Spouses (if individual currently lives in Florida or Texas)
Summary of Assets
Professional Licenses
Property Ownership and value
Vehicle Ownership and value
UCC Lien Filings
Civil Judgments
Bankruptcies

8 According to Microsoft, 'Here are some examples of metadata that can be stored in your documents:
Your name, your initials, or the names of previous document author
Your company or organization name
The name of your computer, or the name of
the network server or hard disk where you saved the document
Other file properties and summary information
Non-visible portions of embedded OLE objects
Document revisions, document versions, template information,
hidden text or cells, and personalized views.'
http://officeupdate.microsoft.com/Articles/Metadata.htm (visited 27 November 2000).

9 http://officeupdate.microsoft.com/Articles/privacy.htm (visited 27 November 2000).

10 By contrast, the French legal system is very protective of personal privacy. See Hauch (1993).

11 Intelligence Identities Protection Act 1982 50 USC 421; Video Privacy Act 18USC 2710; Driver's Privacy Protection Act 18USC 2721.

12 The constitutionality of the statute was upheld by the US Supreme Court (*Reno* v *Condon* 98-1464, decided 12 January 2000).

13 The Intelligence Identities Protection Act 1982 50 USC 421(Sec. 421) makes it an offence intentionally to identify a covert agent of the United States. The offence is punishable by a fine of not more than $50 000 or imprisonment for not more than ten years, or both. The *Australian Security Intelligence Organization Act* 1979 (section 92) makes it an offence to publish the identity of an ASIO officer. The maximum penalty is imprisonment for one year and a fine of $2000 ($10 000 in the case of a body corporate).

14 Responding to the now famous cartoon in the *New Yorker* of 5 July 1993: 61, 'On the Internet nobody knows you're a dog'). Technologies of authentication, including computers incorporating video cameras, make this even more of a reality.

15 Dyson (1998: 277) suggests that consumers are not always skilled at protecting their own interests. See also Rotenberg 1998. Strict advocates of privacy object to its commodification.

CHAPTER 11

The Limits of the Law in Controlling Electronic Theft

The information revolution has given rise to many new ways to commit theft, and many new things to steal. This proliferation of targets and techniques has occurred against a background of increasing recognition that the state is limited in its capacity to control human behaviour, and that security and prosperity in cyberspace will depend on the proper functioning of not just agencies of government but a constellation of institutions and conventions in civil society. While the state may play a part in shaping some of these institutions, or at least in providing a legal framework within which they might operate, others will function more or less independently, to the extent that the role of the state will be at best peripheral. We share Lessig's (1999: x) view of 'a future of control in large part exercised by technologies of commerce, backed by the rule of law'.

Conventional crime has proved a difficult challenge for law enforcement. Electronic theft poses even greater challenges. The three predictors of crime – motivation, opportunity, and the absence of capable guardianship – are not equally amenable to modification. Greed, lust and revenge are fairly robust human characteristics, which do not lend themselves to quick policy fixes. This is not to suggest that efforts to enhance the ethical climate of cyberspace will be misplaced; rather, that they are only likely to be marginally effective, at least in our lifetime. The reduction of opportunities to commit electronic theft, if carried out solely through the curtailment of connectivity, would be counterproductive. Where the economy is not already linked with global forces, it will become increasingly so. 'Pulling the plug' will only impede commerce. Opportunity reduction within the emerging infrastructure of electronic commerce may, however, have much to offer. Capable guardianship is also a viable strategy to adopt for the protection of property in cyberspace. As we have observed, this task is far beyond the capacity of the state alone to achieve.

Technology can significantly enhance guardianship. Denning (1999) describes various technologies for detecting attempted intrusions of information systems. Alarms can indicate when repeated login attempts fail because of incorrect passwords, or when access is sought outside normal working hours. Other anomaly detection devices will identify unusual patterns of system use, including atypical destination and duration of telephone calls, or unusual spending patterns using credit cards.

Guardianship can also be enhanced by market forces. A market is currently emerging for privacy-enhancing technologies to provide secure platforms for electronic commerce, as well as to protect consumers against invasions of privacy. Market forces may also generate second-order controlling influences. As large organizations begin to appreciate their vulnerability to electronic theft or vandalism, they may be expected to insure against potential losses. It is very much in the interests of insurance companies to require appropriate security precautions on the part of their policyholders. Indeed, decisions to set and to price insurance may well depend on security practices of prospective insureds. Subcontractors may also be required to have strict IT integrity programs in place as a condition of doing business. Companies may be able to trade on a reputation for protecting clients' privacy.

Citizen concern about the availability of undesirable content has given rise to the private monitoring and surveillance of cyberspace. Citizen co-production can complement activities undertaken by agencies of the state. An example of collaborative public–private effort in furtherance of controlling electronic theft is the various media available for reporting suspected online fraud.

The policing of terrestrial space is now very much a pluralistic endeavour, and so too is the policing of cyberspace. Responsibilities for the control of computer crime will be similarly shared between agents of the state, information security specialists in the private sector, and the individual user. In cyberspace today, as on terrestrial space two millennia ago, the first line of defence will be self-defence – in other words, minding one's own store.

The chapters of this book proceeded through a discussion of various generic forms of electronic theft, from those on which there is broad consensus over their suitability for the criminal sanction, to those where there is considerable debate over whether they should constitute a public or a private matter. Whatever the case, the prevention and control of electronic theft in the twenty-first century will be a pluralistic endeavour. This concluding chapter acknowledges the limits of the law, and of law enforcement agencies, in controlling electronic theft, and suggests how this burden might be shared between traditional law enforcement bodies, prospective victims of electronic theft, and other institutions of civil society. We will see that legal pluralism is alive and well in cyberspace.

Governments throughout the Western world are encouraging their citizens to take greater responsibility for their own well-being, in areas as diverse as health insurance, retirement income, and crime prevention. Some governments explicitly ask what is a proper state function and what is not, and consciously seek to devolve certain roles downward, and to activate non-state institutions and interests in furtherance of these tasks. In the area of criminal justice generally, Garland (1996: 452) awkwardly but accurately refers to this shifting of the burden of social control as a 'responsibilization strategy'.

Explicitly or implicitly, governments encourage such 'responsibilization' by providing information and/or subsidies to the general public or to non-government actors, by seeking to coordinate the activities of these groups, or by offering to collaborate with them in furtherance of regulatory objectives. Garland cites institutions such as Neighbourhood Watch as an example. In a sense, governments are seeking to engineer a return to a situation that prevailed before the state began to monopolize crime control, at a time when the ordering of society was largely the responsibility of civil institutions.

At first glance, this perspective on the changing role of the state would seem more apposite to conventional criminality than to high-technology crime. In the latter case, governments have been seeking to assert themselves in areas as diverse as content regulation and infrastructure protection. The *Communications Decency Act* 1996 and prohibition on the export of strong cryptography (only recently relaxed) are examples from the United States. In Australia, the *Broadcasting Services Amendment (Online Services) Act* 1999 (Cwlth) seeks to regulate content through the adoption of a complaints mechanism based on industry codes of practice. On the other hand, in many respects the state is following rather than leading, reacting to developments in technology rather than driving them. And the understanding that a great deal of expertise in the social control of electronic behaviour resides outside the public sector means that the state is unlikely to monopolize cyberspace, even if it attempts or pretends to do so.

The Limits of Law

Law, especially criminal law, is often regarded as a panacea for many social problems, but the desire for quick fixes belies the reality that in cyberspace no less than on the ground, legal remedies often have their shortcomings. They may be prohibitively expensive, inefficient, or, at worst, they may actually be counterproductive. As we have seen, legal remedies for the violation of privacy may entail further loss of privacy. Unintended consequences are not, however, limited to the domain of privacy. Attempts at content regulation can inspire the proliferation of 'mirror sites' (sites that replicate or reproduce the content of a site that has been targeted for censorship). Defamation actions may draw even wider attention to unflattering information. Public disclosure of one's vulnerability to fraud, extortion, or other forms of theft may reflect adversely on one's reputation for financial integrity and security, encouraging victims to 'lump it' rather than go public.

The cost of adverse publicity, when weighed against losses already incurred, may be prohibitive. The possibility of recovering assets once dissipated may be low in any event. Moreover, the law can be a cumbersome process – criminal prosecution is a lengthy and often costly undertaking for the state, and the outcome is by no means certain. And even a successful prosecution may result in a fairly lenient sentence.

Another type of problem that arises in the course of criminal prosecution is the need to prove that the offender intended to carry out the proscribed conduct in question. In the case of some frauds involving digital technology, we may ask whether it is possible to prove deception or intention to defraud where the conduct has been carried out against an electronic system as opposed to one directed at human actors. At common law, deception is a matter of the human mind.

Theft perpetrated electronically invariably involves the submission of false or misleading documents or forms, or the making of false statements either orally or in writing. In a digital system, electronic documents and forms are used and statements may be made in digital form. Complex questions of proof and attribution of authorship invariably arise. Not all jurisdictions currently define 'document' or 'statement' to include those in digital media, but many jurisdictions are attempting to satisfy the model law on electronic commerce of the United Nations Commission on International Trade Law (UNCITRAL).

Even if a victim of electronic theft can establish that there has been some non-compliance with laws, the problems associated with taking legal action are considerable and are not limited to the world of electronic commerce. Because many online transactions take place across jurisdictional boundaries, often internationally, the practical difficulties of taking proceedings are ever present: problems of language, time differences, geographical distance if witnesses have to be examined or give evidence in other countries, obtaining legal advice from a foreign country, and the cost and delay involved in communicating internationally.

There are also complex forensic and evidentiary problems associated with investigating online illegality. These include difficulties associated with collecting evidence, as computer files may have been disguised or destroyed, leaving no traces of the illegality concerned. Files may also be encrypted, making them unavailable for legal purposes. Presenting evidence in court can also be difficult as a result of exclusion through the use of hearsay and best-evidence rules to electronic documents where other evidence may not be attainable. Many jurisdictions are currently attempting to overcome these limitations through the adoption of electronic evidence acts.

For many of these reasons, in addition to the formidable challenge of proving guilt beyond reasonable doubt, some authorities will often opt for the expeditious neutralization of an offender by means of injunctive remedies or consent orders requiring the offender to cease the objectionable activity, rather than chance the cost, delay and uncertainty of a criminal prosecution. One such instance was detailed in Chapter 5, where we saw how Telstra, the large Australian carrier, may on occasion opt to bill the perpetrator of theft of services for the full amount of the misappropriation rather than prosecute. Others, discussed in Chapter 6, concerned cases where the immediate cessation of sharemarket fraud took precedence over potential criminal sanctions.

The Limits of Law Enforcement

The decision to mobilize the law primarily belongs to the victim. Just as the victim of a sexual assault may decide that calling the incident to the attention of the authorities may lead to substantial inconvenience and further psychological trauma, so too the victim of electronic theft may conclude that the immediate financial loss would be outweighed by the damage to his commercial reputation should his misfortune become public knowledge. So it is that an unknown but presumably very large proportion of computer-related crime never reaches police attention in the first place. Sometimes victims might not even realize that they have been defrauded.

As well as this, the capacity of the police to respond to electronic theft is often manifestly inadequate. Throughout the Western world, police are being asked to do more with less. 'Street crime', by which we mean conventional crimes against the person and property, continues to be enough of a problem in most Western industrial societies and commands the largest slice of the law enforcement pie. This problem is compounded by a lack of staff. Law enforcement officers who develop skills in forensic computing can often double their salaries in the private sector, and many leave the police for greener pastures, contributing to a significant brain drain. Training and retaining police who are competent computer crime investigators is a challenge for law enforcement agencies everywhere. One instance of the types of problems that arise as a result of inadequate staffing was noted by Richtel (1999), who reported that law enforcement authorities in the United States currently had a backlog of confiscated computers awaiting content analysis.

Improvement for the current staffing situation does not appear imminent. Although an increase in qualified graduates may enhance the pool from which forensic experts may be drawn, police everywhere must still decide on which incidents to investigate. This is typically decided on a triage system, the result of which is that digital crime is invariably afforded less importance because of the cost and effort required to complete a successful investigation and prosecution.

New Solutions

In recent times it has become common to distinguish between command and market as alternative vehicles for achieving social objectives. A good deal of discourse, at least that emanating from fundamentalists of the Left or Right, implies that they are mutually exclusive.

It would seem, however, that this focus on ideal types is misconceived, as it is not always possible, or even desirable, to differentiate precisely between public and private. Just as Ayres and Braithwaite (1992) observe that the regulation/deregulation dichotomy has long outlived whatever usefulness it may have had, it is no longer very useful to speak in terms of government versus private institutions. Instruments of public policy appear increasingly to involve a blend of public and private resources. Markets may themselves be constituted, and their viability preserved, by government regulation (Shearing 1993).

Foucault (1977) described modern society as a proliferation of disciplinary networks, a 'carceral archipelago' of government and private institutions of social control, which enables governments to exercise power at minimal economic cost, with minimal visibility, and in a manner likely to provoke minimal resistance. Indirect institutions of social control may be more effective and at the same time be perceived as more legitimate than the conventional exercise of brute force.

To some, this might appear sinister. But in some regulatory spheres, it need not be so. On the contrary, Galanter and Luban (1993) see in this the potential for a decentralized and more participatory regulatory order. Also in this vein, Ayres and Braithwaite (1992) have pointed towards a regulatory republicanism, where an enlightened private sector and an informed public, through deliberation and constructive participation, can contribute productively to the regulatory process. More recently, Braithwaite and Drahos (2000) call for active world citizenship in achieving global regulation. If, to paraphrase Foucault, the hands of the state are incapable of controlling everything, how then may the state most constructively enlist the energies of private interests and the general citizenry in furthering security and prosperity in cyberspace?

Legal Responses

Although the flexibility and adaptability of the common law to deal with changing social conditions represents one of its great advantages, sometimes effective legal changes take a considerable time to come about. The most innovative scientific developments occasionally leave the law in a state of confusion, requiring considerable effort on the part of law reform agencies and legislatures to respond effectively. The introduction of electricity, as we have already seen, is a case in point. Originally, electricity, an intangible form of property, could not be stolen, as larceny could not be committed in respect of something incapable of being physically removed. Parliament was required to fill the gap by creating statutory offences of theft of electricity which could be applied, for example, to those who illegally obtained telecommunications services fraudulently (see Grabosky and Smith 1998: 66).

More recently, legislatures throughout the world have begun to address the inadequacies found to exist in laws that govern transactions carried out electronically. In the context of crimes of acquisition, amending provisions have been inserted into existing statutes relating to larceny. We saw this in the case of subsection 3 of section 63 of Australia's *Crimes Act* 1914 (Cwlth) (computer-generated forgery) and also in the Australian Model Criminal Code, which has proposed a variety of changes to the law relating to fraud and forgery throughout Australia (see Criminal Code Amendment [Theft, Fraud, Bribery and Related Offences] Bill 1999 [Cwlth]; Model Criminal Code Officers Committee 1995, 2000). The definition of 'document' in Clause 19.1(1) relating to forgery and offences involving false documents includes 'a disc, tape or other article from which sounds, images or messages are capable of being reproduced', while Clause 19.1(2) provides that 'a reference to inducing a person to accept a false document as genuine includes a reference to causing a machine to respond to the document as if it were a genuine document'.

The rapidity with which changes occur in the digital age, however, makes the conventional processes of law reform cumbersome. Introducing a legislative change to accommodate the latest development in computer user authentication technologies, for example, may take years, by which time the technological development in question may be outmoded and the reform no longer applicable.

This has given rise to the need for legislative responses in the digital age to be written in 'technology-neutral' terms that are unlikely to be affected by novel technological changes in the short term. Recent reforms that seek to accommodate electronic commerce have been created with this concern in mind. The Model Law on Electronic Commerce produced by UNCITRAL in 1996 is a case in point. The Model Law sets out recommended legal provisions to remove any legal obstacles to the development of electronic commerce such as provisions capable of governing the recognition of electronic signatures and the admissibility of computer evidence. The UNCITRAL Model Law has been adopted in full or in large part by many jurisdictions worldwide including Singapore, much of Europe, and several jurisdictions in the United States. Australia has adopted tracts of the law in the *Electronic Transactions Act* 1999 (Cwlth). The significance of such laws is that they provide that electronic communications be treated in the same manner as paper-based communications save in specific areas that are especially prone to fraud (such as testamentary documents). The target of such laws is to remove uncertainty in the civil arena, but it may be argued that this should be extended to the criminal law. It will remain to be seen whether or not such laws will in fact continue to apply when the state-of-the-art technologies of the year 2000 have become a distant memory.

The other difficulty created by digital technologies is that commerce in both its legitimate and illegitimate forms invariably crosses geographical and jurisdictional boundaries. This requires the international compatibility of legislative responses. Already we have seen some achievements in this regard.

Beginning in the late 1980s the Council of Europe has taken significant steps towards the harmonization of laws relating to computer-related crime. A Draft Convention on Cyber-Crime, under development at the end of 2000, contained general sections on the substantive criminal law, search and seizure of electronic data, jurisdiction, and mutual assistance. The Council recommended measures to criminalize the following: illegal access to computer systems; illegal interception of or interference with data; the production, sale or procurement of 'hacking' tools or software; computer-related fraud and forgery; offences relating to child pornography; and offences relating to copyright infringement (Council of Europe 2000).

Although much can be achieved by the introduction of internationally uniform and technology-neutral laws, difficulties will invariably arise in applying such laws to the circumstances created by new technological developments. Unfortunately, the activities of some cybercriminals will escape sanction in the short term. With effective dialogue and informed cooperative action, however, some successful prosecutions may be achieved. It is to be hoped that a few well-publicized cases will enable at least some of the community to be deterred from engaging in electronic theft.

Self-regulation through Enforceable Codes of Conduct

An alternative approach seeks to rely on industry-based self-regulation. This may be voluntary, or backed by legislative enforcement provisions. The device being adopted in a number of countries for the regulation of online conduct is for codes of conduct to be created that specify acceptable forms of behaviour (as agreed internationally) and for these to be made enforceable through local legislative means. Such codes of conduct provide not only a widely disseminated statement of existing laws and acceptable practices that help to create a culture of compliance within specific industries, but also often include dispute resolution procedures and sanctions for non-compliance with the rules in question. Although many countries now have codes of conduct to regulate various online activities, the following discussion will focus primarily on Australia, which has been a world leader in this area, particularly in codifying desirable practices on the Internet.

Marketing codes

In Australia, the marketing and media industries have established codes of conduct that target particularly vulnerable groups of consumers such as children, as well as specific content such as obscene materials, and products such as therapeutic goods, tobacco and alcohol. The Media Council of Australia, for example, administers a variety of voluntary codes of practice relating to the advertising of therapeutic goods, slimming products, alcohol and tobacco products (see Pearson 1996).

In Australia, the ACCC has developed Guidelines for Advertisers, while state and territory departments of consumer affairs also have guidelines on compliance with local laws such as those relating to the protection of privacy (see below). Some industry groups also have their own codes of practice such as the Australian Publishers' Bureau Advertising Code of Practice, which sets out its requirements for acceptable advertising in six short paragraphs. These codes and guidelines reflect the provisions of the general law and do not take away existing consumer rights. They also often operate across jurisdictional boundaries, which increases the potential for the development of uniform practices.

In December 1997, the Australian Ministerial Council on Consumer Affairs released the Direct Marketing Model Code of Conduct to regulate the conduct of those involved in the direct-selling industry. The Code is administered by the Australian Direct Marketing Association (ADMA), which was established in 1966 as the peak industry body for companies and individuals engaged in direct marketing in Australia, and applies to telemarketing, mail-order and Internet sales.

All ADMA members must undertake to abide by the voluntary Direct Marketing Code of Practice published by the Association, which seeks to ensure that direct marketing by members complies with the highest standards of integrity. The 'Standards of Fair Conduct' within the Code govern the making of an offer, identification of the advertiser, the use of incentives, the placing of orders, fulfilment of orders and the use of mailing and telephone lists. Arrangements are also made for the arbitration of disputes and members agree to comply with all legal requirements.

The Code also specifically refers to direct marketing carried on electronically. It states, for example:

> Clear, complete and current information about the identity of businesses engaged in electronic commerce and about the goods and/or services they offer, should be provided to customers. Additional information should be provided to address particular aspects of digitised goods and services, such as technical requirements or transmission details (cl. D2).

Of particular concern is the problem of online advertising directed at children, who are often attracted to advertising material and are likely to be misled. Some responsible businesses are aware of the problem, such as Motorola, which sells mobile telephones and whose site contains the following message:

> Motorola products and services are directed at an adult market, and therefore this site is intended for use by adults only. Motorola encourages parents to take an active role in their children's use of the Internet, and to inform them of the dangers of providing information about themselves over the Internet. No information should be submitted to or posted on this Web site by users under 13 years of age without the consent of their parent or guardian. Users that Motorola knows to be under the age of 13 are required to provide the e-mail address of their parent or guardian so that Motorola may alert the parent or guardian of their child's use of this website. The parent or guardian must consent to Motorola's

> collection of their child's personal information. A known child user will be restricted from providing personal information until such consent is received. No information collected from users known to be under age 13 will be used for any marketing or promotional purposes outside Motorola, Inc. (Motorola 2000b)

Failure to comply with the code may result in members' conduct being investigated by a Code Authority established by the Association. Sanctions include orders requiring members to take remedial action or to give an undertaking not to repeat the breach of the code, to issue a formal written admonition and/or to publish that admonition for serious breaches of the code, or to recommend revocation of membership.

In Britain, the New Media Council was set up by the Direct Marketing Association to act as a forum for those who market across non-broadcast electronic media, including the Internet. British Codes of Advertising and Sales Promotion Codes from the Advertising Standards Authority, Direct Marketing Association and Independent Committee for the supervision of Standards of Telephone Information Services all provide specific guidelines on acceptable conduct. With such a proliferation of guidelines, part of the challenge lies in establishing which organization is the most appropriate to be involved in any specific case.

The possible application of the Control of Misleading Advertisements Regulations 1998 (Eng.) to Internet trading is currently under debate in Britain. Internet service providers argue that, like a telecommunications or postal company, they are providing the medium but should not be held responsible for the messages people transmit.

Content regulation codes

Codes of conduct have also been developed to regulate online content. This has mainly taken place with respect to obscene and offensive material, although the possibility exists that such regulatory measures could be extended to misleading and deceptive online content as well. The Committee of Australian University Directors of Information Technology (CAUDIT 1997), for example, has a Code of Practice Relating to Content That May Infringe Censorship Laws. The code is intended to provide guidance to university directors of information technology regarding the regulation of the behaviour of staff and student users in the context of public networks.

The Internet Industry Association recognizes that the Internet should provide a means of controlling access to content while acknowledging that it is impractical to filter all Internet content. Accordingly, it endorses methods by which content can be recognized and possibly excluded by content filter technologies as the most practical means of empowering responsible adults to control access to the Internet to determine appropriate controls on content.

Questions of freedom of speech and the practicality of regulating content are, however, the major areas of concern in adopting content regulatory approaches. In 1999, the Australian Federal Government enacted legislation that established a co-regulatory scheme to address risks associated with illegal content and with content that is unsuitable for children (see Grainger 1999). The *Broadcasting Services Amendment Act* 1999 makes use of a range of regulatory responses including a complaints hotline and industry codes of practice. In effect, the Act treats material disseminated through the Internet in a similar manner to material broadcast on public television networks.

Action to be taken in relation to prohibited content that is subject to complaint differs depending on whether the content is hosted in Australia or overseas. If it is hosted in Australia, the Australian Broadcasting Authority (ABA) is required to issue take-down notices to the content host. If it is hosted overseas, the ABA will notify ISPs who are to take action in accordance with their codes of practice. If it is hosted overseas and is also sufficiently serious (e.g. illegal material such as child pornography), the ABA will refer the material to the appropriate law enforcement agency (Grainger 1999).

Non-compliance with the legislation is made a criminal offence carrying penalties of a fine of up to A$5000 for an individual or A$27 500 for a corporation, with penalties accruing for each day the contravention continues. The regulatory scheme applies to the activities of ISPs and Internet content hosts only. The government has stated that it will be encouraging the states and territories to develop uniform legislation that will complement the Commonwealth legislation and cover the activities of users and content creators (Grainger 1999).

Arguably a similar regulatory regime could be introduced to deal with deceptive and misleading advertising content on the Internet which emanates from local hosts. Although this may have the effect of reducing illegal and fraudulent material that is generated from one's own country, it would have no effect on foreign-based content, which would require regulation by each host jurisdiction. The global nature of cyberspace means that content regulation is likely to remain an elusive objective.

Privacy codes

In another initiative, the Australian Government has introduced legislation to protect individual privacy from inappropriate access or disclosures by private sector entities. Traditionally, as we have seen in Chapter 10, privacy legislation has generally been restricted to the protection of personal information provided to public sector agencies. In Australia, for example, information on income taxation, consumer credit, spent criminal convictions, health, and certain other matters is regulated by legislation. The *Privacy Amendment (Private Sector) Bill* 2000 (Cwlth), introduced into the Australian Federal Parliament on 12 April 2000, aims to establish a single comprehensive national scheme for the appropriate collection, holding, use, correction, disclosure, and transfer of personal information by private sector organizations. This legislation was introduced to enable Australia to meet its international obligations relating to privacy and recognizes that important human rights and social interests compete with privacy, including the general desirability of a free flow of information (through the media and otherwise) and the right of business to achieve its objectives efficiently (Privacy Commissioner 2000).

The scheme adopted makes use of legislative backing for industry codes of conduct that have been developed within the private sector in consultation with the federal Privacy Commissioner. In the online environment, the new legislation would require website operators who collect personal information online to take reasonable steps to ensure that Internet users know who is collecting their information and how it is used, stored, and disclosed. The legislation would also allow people to gain access to their records and to correct those records if they are wrong. Under the proposed legislation, organizations would have to protect people from unauthorized access and disclosure of personal information that they hold, and website operators who handle personal information would have to address issues of data security such as encryption. The proposed legislation would also require organizations to make public their policy on privacy, which means in practice that all websites would have to include a clearly identified privacy statement (Privacy Commissioner 2000).

The Bill would also apply to direct marketing by electronic mail. Although it would permit the use of personal information for direct marketing purposes, individuals would have to be given the opportunity to opt out of receiving any further direct marketing material. This would regulate spamming from Australian private sector organizations. Although the proposed legislation provides benchmarks, private sector organizations would be able to adopt higher standards such as those proposed by the Internet Industry Association.

In order to deal with the problem of transnational activities, the Privacy Amendment (Private Sector) Bill would apply in some circumstances outside Australia to ensure that organizations do not avoid privacy protections simply by moving information overseas. The Bill draws a distinction between Australian and foreign organizations and provides that where an Australian organization deals with information about Australians, the Bill would apply both inside and outside Australia. For example, if an Australian company collects and stores information about Australians overseas, the company would have to apply the safeguards set out in the Bill.

In order to regulate the behaviour of foreign organizations operating outside Australia, it would be necessary for that organization to establish a strong jurisdictional link with Australia. In the Bill, the link would be based on a range of factors including the requirement that the foreign organization carry on business in Australia and deal with information about Australians, and that the information must have been collected, or held at some time, in Australia. Where, for example, a foreign company collects information about Australians in Australia and then moves that information overseas, the company would have to apply the safeguards set out in the Bill. In general, Australian-based organizations would also have to ensure that comparable privacy safeguards applied before transferring information to overseas organizations.

Internet service provider codes

Internet service providers have established a number of industry-based organizations, several of which have their own codes of conduct. In Australia, the Internet Industry Association, for example, has a code of practice (Version 6.0 as approved by the ABA on 16 December 1999) that was based on generally accepted international standards, such as Australian Standard AS-4269-1995, a wide range of existing and related codes, the Ministerial Council on Consumer Affairs' Guide to Fair Trading Codes of Conduct, and various regulatory schemes in related industries (Internet Industry Association 1999). A number of local groups also have Codes such as the South Australian Internet Association's Code of Ethics and Conduct and the Western Australian Internet Association's Code of Conduct.

Non-compliance with voluntary codes of conduct is an area of continuing concern because some organizations face few, if any, consequences for failure to comply with the provisions of their code. Occasionally consequences of non-compliance are provided for, such as in the *Fair Trading Act* 1987 (NSW), section 74. This provides that failure to comply with an industry code of practice may result in an undertaking to discontinue the offending conduct, to comply with the code in the future, or to take action to rectify the consequences of the contravention. Stronger consequences may be suspension or disqualification of the offending person from continuing membership of the industry group in question, which could entail substantial financial repercussions.

The principal limitation of codes of conduct as a regulatory mechanism is that their operation is limited to those who have agreed to comply with their provisions. Although this may be adequate for large organizations such as those that regulate direct marketing, the controls are usually restricted to specific geographical regions. In the world of online marketing and advertising, where information travels so easily across borders, the possibility of consumers or investors being misled by information from some overseas source is greatly increased. Conflicting guidelines may also be established in different countries to regulate essentially similar activities.

In the global marketplace, international codes of conduct will ideally be agreed on by groups representing essentially similar interests. One could imagine, for example, an international code of conduct that would apply to all ISPs throughout the world. OECD countries, for example, are aware of this need and have already begun efforts to create of a set of guidelines for comparable self-regulatory regimes in different OECD countries (Bridgeman 1997). The problem with uniform international codes of conduct is the accommodation of local public sentiment in setting standards. In regulating obscene and objectionable content, for example, this has proved to be a considerable hurdle (Butler 1996). In the field of misleading and deceptive practices, however, international consensus might be more easily achieved.

Preventive Strategies

A wide range of strategies have been developed to prevent electronic theft. In theoretical terms, their function is that of guardianship. These range from the creation of guidelines and policies to the use of computer-based security techniques. Some take the form of regular surveillance of the Internet in order to locate objectionable and illegal practices, providing educational material warning users of risky schemes; others use authentication technologies to permit individuals to know with certainty who they are dealing with in the online world. Caution is needed in relying on some preventive strategies, however, as they may infringe rights of privacy, liberty and free speech.

Institutional/organizational

Online monitoring and investigation by regulatory authorities

In order to address some of the forensic and evidentiary problems noted above, regulatory and law enforcement agencies are beginning to establish in-house specialist units to detect and investigate online illegality. In Australia, in April 1999, ASIC established an Electronic Enforcement Unit specifically to deal with online business practices. In the United States, the FBI has created an Internet Fraud Complaint Centre to deal with consumers who have been defrauded electronically.

Consumer and investor protection agencies around the world recognize that the Internet has no geographical boundaries and are working together to combat fraudulent activity. The ACCC, its US counterpart the FTC and Portugal's Instituto do Comidor recently worked together to stop a 'page-jacking' scam that led unsuspecting users into a series of pornography sites. Consumer protection agencies in Canada, Australia and the United States agreed in July 2000 to establish an ongoing system for information sharing in relation to cyberfraud.

Risk management

Most public and private sector agencies throughout the world have detailed policies in place which provide guidelines on the establishment, implementation and management of measures to reduce the risks of economic crime. There has also been a recognition in recent times of the need to create an ethical environment in workplaces by educating staff about the desirability of complying with laws and codes of practice. Policies also need to deal with the specific online behaviour of employees. Many organizations establish guidelines, for example, on access to and use of the Internet for private purposes, personal use of electronic mail, downloading organizational software, and the use of copyright material.

Personnel monitoring

One of the most important areas in which digital fraud can be curtailed lies in ensuring that trustworthy and reliable staff are employed, particularly in senior positions of responsibility. The administration of modern technologically based security systems involves a wide range of personnel, from those engaged in the manufacture of security devices to those who maintain sensitive information on passwords and account records. Each has the ability to make use of confidential information or facilities to commit fraud or, more commonly, to collude with people outside the organization to perpetrate an offence. Preventing such activities requires an application of effective risk management procedures within agencies which extend from pre-employment screening of staff to regular monitoring of the workplace. Long-term employees who have acquired considerable knowledge of an organization's security procedures should be particularly monitored, as it is they who have the greatest knowledge of the opportunities that exist for fraud and the influence to carry them out.

Computer usage monitoring

Employees' computer use and online activities can be monitored by software that logs use and allows managers to know, for example, whether staff have been using the Internet for non-work-related activities. Ideally, agreed procedures and rules will be established that enable staff to know precisely the extent to which they may use office computers for private activities, if at all. If organizations do permit staff to use computers for private purposes, then procedures should be in place to protect privacy and confidentiality of communications, subject, of course, to employees obeying the law.

Technical safeguards

The greatest safeguards against electronic theft will probably be technologies of prevention. From fingerprint scanners that reduce the opportunity for social security fraud, to digital watermarks that discourage the pirating of copyright material, to access controls that make industrial espionage much more difficult, to the various privacy-enhancing technologies that facilitate the safeguarding of personal information, some of the most promising fixes will be technological.

Where certain online activities have been prohibited, many organizations now monitor the activities of their employees, sometimes covertly through video surveillance or by checking electronic mail and files stored on servers. Filtering software may also be used to prevent staff from engaging in certain behaviours. 'Surfwatch', for example, can be customized to deny employees access to specified content. When the employee requests a site, the software matches the user's ID with the content allowable for the assigned category, then either loads the requested page or advises the user that the request has been denied. The software also logs denied requests for later inspection by management.

The use of computer software to monitor the business activities of government agencies also provides an effective means of detecting fraud and deterring individuals from acting illegally. The Australian Health Insurance Commission, for example, employs artificial neural networks to detect inappropriate claims made by health care providers and members of the public in respect of various government-funded health services and benefits. In 1997/98, this technology contributed to the HIC locating $7.6 million in benefits that were paid incorrectly to providers and the public (HIC 1998).

In addition, revenue authorities can make use of information derived from financial transaction-reporting requirements to identify suspicious patterns of cash transactions that could involve illegality or money laundering. In Australia, in 1997/98, the Australian Taxation Office attributed more than $47 million in revenue assessed to its direct use of information provided by AUSTRAC. In one case, a taxpayer and associated entities had transferred more than $1.3 million to a tax haven. After an investigation, more than $6 million in undeclared income was detected (AUSTRAC 1999).

Technological solutions, however, entail implementation and use by individuals who need to be instructed in how to make appropriate and efficient use of the technologies they seek to adopt. The considerable benefits that may arise through the use of digital signatures to protect financial communications, for example, could be rendered nugatory simply through users being negligent in their use of passwords when they begin using such technologies. Technology, therefore, will only be effective if those who make use of it are appropriately trained and instructed in its correct use.

User authentication

Authentication of one's identity is a case in point (see Chapter 4). At present, most authentication procedures involve the use of passwords or PINs. Ensuring that these are used carefully and are not compromised represents a fundamental fraud control measure. In addition to user education, a variety of innovative ideas have been developed to protect passwords and to enhance user authentication (see Alexander 1995).

Systems are available that change passwords regularly, or that deny access after a specified number of consecutive tries using invalid passwords. Terminals have been devised with automatic shutdown facilities that operate when they have not been used for specified periods. Single-use passwords, challenge-response protocols, and call-back systems have also been devised as a means of carrying out user authentication.

Finally, space geodetic methods have been devised to authenticate the physical locations of users. In the future, many user authentication systems will make use of so-called biometric identifiers which make use of an individual's unique physical characteristics.

Those who manufacture computer hardware and software have a role to play in protecting themselves as well as their customers from electronic theft. Basic principles of target hardening may help to ensure that hardware cannot be stolen and also that security devices are available and easy to use. Keyboards and 'mice' which contain inbuilt biometric access devices such as a fingerprint scanner are now available. Devising software that cannot be pirated seems a less complex solution to copyright infringement than trying to locate infringing copies and prosecute those thought to be responsible.

Pluralistic Approaches

In a very perceptive article published over a decade ago, the sociologist Gary Marx (1987) observed five trends in investigations:

- joint public-private investigations
- public agents hiring or delegating authority to private police
- private interests hiring public police
- new organizational forms in which the distinction between public and private is blurred
- the circulation of personnel between the public and private sector.

While Marx was referring to terrestrial policing, and specifically to undercover operations, his perception of trends would appear no less applicable to the investigation of crime in cyberspace. Specifically, the challenges confronting law enforcement agencies, in light of the constraints discussed above, are requiring a degree of resourcefulness that is unprecedented in modern policing.

Among other things, this will entail a range of relationships with a diverse set of organizations and individuals. First among these are the victims, who are very much the gatekeepers for computer crime control. Without information from victims, be they companies, government agencies, or private citizens, we will remain in the dark about the incidence and prevalence of electronic theft, and about the modus operandi of electronic thieves. It is important to encourage victims to come forward and share their experiences with an authority able to analyse these data for purposes of intelligence and enforcement.

Of particular importance is the need to develop appropriate guidelines on reporting computer misuse and abuse. Many jurisdictions now have public interest disclosure legislation which aims to ensure that those who report illegal activity are not disadvantaged by their conduct. In the case of computer-based illegality, as in other areas of crime, severe penalties could be imposed on individuals who engage in or attempt or conspire with others to carry out acts of reprisal against those who disclose illegality in the public interest. To date such remedies have rarely been used.

Second, those with assets to protect must recognize that responsibility for prevention always begins at home. Basic practices of residential security, such as the installation of locks and burglar alarms, are feasible; putting a police officer on every street corner is not. The same principle applies in cyberspace. Information security must be the concern of everyone.

The third principle is the necessity of partnerships and strategic alliances. These must occur between government agencies within jurisdictions, between agencies across jurisdictions, and between institutions in the public and private sectors. Given the global nature of cyberspace, securities regulators around the world must cooperate in the detection and suppression of market-related illegalities. The versatility of some cybercriminals and the similarity of some offences means that agencies which oversee different regulatory domains must also cooperate. And the pluralization of expertise in information security means that agencies of the state will increasingly call on expertise resident in the private sector to help in controlling cybercrime. In some respects this is hardly new; the pursuit and capture of Kevin Mitnick in 1995 was a joint effort involving law enforcement, telco staff and IT security professionals (Shimomura 1996).

Telecommunications carriers

In many countries telecommunications carriers cooperate regularly with law enforcement agencies to facilitate the identification of online offenders and to assist in their prosecution. In Australia, legislation specifically requires carriers and service providers to take steps to ensure that their facilities are not used for illegal purposes, and to facilitate law enforcement investigations, subject to basic protections such as those governing communications interception.

In the United States, where obvious fraud prevention steps have not always been taken by all carriers, some subscribers have argued in civil proceedings that they should not be held personally liable for fees incurred by reason of telecommunications fraud where the carrier has acted negligently in failing to ensure that systems operate securely.

Service providers

Some service providers require, as a condition of a service contract, that users undertake to observe the law. In addition, providers may monitor certain areas of their service, such as chat rooms, and disconnect the services of those who may have violated the terms of their contract. In some cases, governments may 'conscript' service providers by requiring them to report suspected illegality to law enforcement authorities.

Financial institutions and other large organizations with a significant IT capacity are those interests that arguably stand to benefit most from security in cyberspace. The most successful of these will have invested substantially in information security and assembled substantial expertise. Private sector providers of forensic computing services, ranging from the smaller, more specialized companies to divisions of large accounting firms, are able to assist individual clients, as well as enforcement agencies.

IT enthusiasts, including the relatively ethical members of the hacking community, often command formidable expertise in information security which they may be willing to share with clients, on a fee-for-service or *pro bono* basis. Other institutions, external to law enforcement, will also be in a position to contribute to a more secure environment in cyberspace. These range from computer emergency response teams or CERTS, which provide assistance in the event of a mishap, to a variety of other non-government organizations.

The International Chamber of Commerce (ICC) has opened a new division to assist 7000 member companies around the world to protect themselves against computer-related crime. In addition to identifying how and where attacks occur, the Chamber seeks to give security information to members. The new ICC cybercrime unit is developing a database on criminal activities in cyberspace, and will facilitate information exchange between the private sector and law enforcement (International Chamber of Commerce 2000)

In 1999 the RAND Corporation sought to establish a cyber assurance program to assist participating organizations in managing risk to their information systems. In addition, RAND proposed publishing online newsletters relating to information security, hacking and computer-related crime. They were intended to deliver expert guidance and analysis about existing and developing threats from, and defences against, cybercrime, cyberterrorism and cyberwarfare, for the combined business-government market, and to provide coverage of new research and its implications, analysis of trends, and alerts when appropriate.

One letter was to have been aimed at corporations that develop, manufacture and sell cybersecurity products and services, and to organizations that purchase and use such systems. It was to address existing and developing cybersecurity methods, strategies, systems and applications; coverage of new research and its implications; and trend analyses.

Another webletter was to have covered the activities of hackers and crackers around the world that affect business and government systems. It was targeted to businesses and government agencies vulnerable to threats by hackers and crackers. It was intended to deliver expert guidance and analysis including a listing of significant hacker/cracker incidents, how they are being defeated and defended against, trends in the hacker/cracker community, and rumours and upcoming threats compiled by monitoring the hacker/cracker bulletin boards on the Net and via other means. The program floundered, however, because of apparent lack of interest on the part of prospective subscribers (Peter Greenwood, pers. comm. 2000).

Private Civil Litigation and Prosecution

There are a number of areas outside cyberspace where the enforcement efforts of government are complemented, if not entirely eclipsed, by those of non-government institutions. One might create certain specified rights, confer them on private parties, and leave it to those private parties to enforce the rights in question. Many systems of patent, trademark and copyright already rely on such private enforcement. In Australia, the vast majority of actions brought under the *Trade Practices Act* 1974 are brought by private parties. Companies and securities regulators also encourage private actions on the part of aggrieved parties. Successful private litigation may result in damage awards that exceed civil or criminal penalties available to the state; the deterrent effect posed by many potential private enforcers should not be underestimated.

A second avenue of private enforcement entails empowering third parties to undertake enforcement actions on behalf of the state. The principle is by no means a modern one. A fourteenth-century English statute specified that 25 per cent of fines imposed on stallholders engaged in trade after the close of a fair be paid to citizens intervening on behalf of the King (5 Edw III, Ch. 5 [1331]; Boyer and Meidinger 1985: 948).[1]

Private enforcement in the public interest is alive and well in many jurisdictions. In Britain, the generally available right of private prosecution has been exercised in the public interest by third parties. Each year, the Royal Society for the Prevention of Cruelty to Animals investigates tens of thousands of complaints and successfully prosecutes about 2000 cases (Harlow and Rawlings 1992). Short of actually carrying a case to prosecution, private institutions often play a leading, if not exclusive, role in the investigative process. As we have noted, copyright enforcement is already a significantly private matter. In English-speaking democracies, many cases of suspected insurance fraud are investigated by employees of the insurance industry, many of whom have previously served in law enforcement and who thereby command substantial expertise in investigation.

The main problems associated with relying on private remedies relate to the cost and complexity of the legal system. Although large corporations may be willing and able to pursue remedies through the courts, it is rarely the case that individuals are able to take action unless the stakes are high in personal or financial terms. Where consumer rights have been infringed, however, and where individual losses may be small, the possibility arises of class actions being taken to deal with practices that have more general application.

'Counter-hacking'

Although some victims may be tempted to take the law into their own hands by 'counter-hacking' an offending individual or website, one may never be absolutely confident that that the target is in fact the offender. By adopting the electronic identity of an innocent party, an offender may elicit an attack that can lead to the damage or destruction of that party's assets. Such collateral damage can be significant, and can get one into serious trouble.

Whatever the relative roles of the public and private sectors in policing cyberspace, there can be little doubt that an increasing amount of traffic on the information superhighway will occur across international frontiers. This will be the case with illicit activity as well as with legitimate commerce. The globalization of electronic commerce, and of electronic crime, will necessitate even newer forms of cooperation between public law enforcement agencies. The impediments to enforcement across sovereign frontiers will be that much greater in the absence of consistent laws which provide the basis for mutual assistance in criminal matters, including investigation and extradition. In some cases of dual criminality, cost-effective arrangements of prosecution in the defendant's home jurisdiction may be a viable alternative to extradition.

Summary of the Forms of Electronic Theft

Let us briefly revisit the various forms of electronic theft which were discussed in Chapters 2 to 10, with particular emphasis on the institutional arrangements and technologies that exist for their prevention and control.

Internet payment systems

Merchants and governments who carry out transactions electronically are well placed to protect themselves against illegal online conduct. Knowing the true identity of customers and ensuring adequate credit-worthiness are obvious first steps, and technologies can greatly assist in these tasks. It is also essential to have appropriately secured systems in place to transfer funds electronically. Merchants also need to take steps to ensure that trustworthy staff are employed, and that those who manage computer systems are monitored in terms of their integrity.

The risk of electronic fraud against financial institutions may require more stringent defences, even if these deter some customers or decrease business efficiency. The area of user authentication is a case in point. At present, many financial institutions are happy to live with the risks associated with manipulation of passwords and PINs and are reluctant to try such solutions as plastic cards incorporating photographs or biometric authentication technologies. As online banking increases, however, customers may demand greater security and be more willing to accept any perceived problems that enhanced user authentication systems may entail.

Financial institutions may also become more willing to monitor transactions on behalf of customers. In Canada, for example, older people have begun authorizing their banks to monitor their accounts in order to discover unusually large or anomalous transactions. The bank is then authorized to raise its concerns with the account-holder and to warn of the possibility of fraud. Account-holders, however, retain full rights over their accounts and may elect to disregard any warnings given. This scheme has already resulted in one older person being prevented from losing Can$20 000 through telemarketing fraud after an earlier incident in which Can$40 000 had been lost (Zanin 1998). Similarly, in Massachusetts, a program in which bank employees received special training in the identification of possible cases of abuse of older people's bank accounts led to the identification of a number of cases of financial abuse (Price and Fox 1997).

In another initiative, believed to be a world first, the Australian Payments System Council, which has only recently been dissolved, conducted a review of the security standards evident in retail electronic funds transfer systems in 1990–91. Where shortcomings were revealed, these were followed up at both institution and industry level, and more general guidelines published to improve EFT security. The Council also conducted an annual review of the level of compliance with various financial system codes of conduct (Australian Payments System Council 1998).

In order to solve the problem of consumers and merchants misrepresenting their identities in online transactions, electronic user authentication systems are starting to be used. Public key systems that use encrypted data transmissions are one way of helping consumers and merchants to be confident of the identity of the person they are dealing with. Such technologies would not prevent people from illegally getting access to private cryptographic keys by stealing tokens that hold keys or by presenting fabricated documentation (see Office of Government Information Technology 1998), but they do represent a much more secure way of conducting online transactions than simply by trusting material that is displayed on the Internet and hoping it will be secure.

Extortion

For large organizations, defence against extortion will be similar in many respects to defence against EFT fraud. A comprehensive system of security should be in place, embracing personnel management, defences against unauthorized intrusion, and the identification of anomalous patterns of use. There should also be contingency plans in the event of an extortion threat.

Since individuals lack the infrastructure of large organizations, extortion threats against them will entail different responses. Notification of both law enforcement and one's own service provider should be the first response. In the event of persistent abuse, engagement of private forensic computing services may be appropriate.

Fraud against government

Governments will defend themselves against fraud by having in place traditional security arrangements, especially technologies of access control and authentication. In the future, fraud control in the public sector will become largely an electronically focused activity. This will entail a combination of biometric authentication and automatic anomaly detection. Nevertheless, because the most skilful perpetrators of fraud strive to be as unobtrusive as possible, a degree of low-tech surveillance and traditional investigation will still be appropriate.

Theft of services

Here again, the solutions will mainly involve applications of technology. The challenge is one of risk management, whereby an optimal solution will prevail. Constraints on criminal opportunities will be introduced to the extent that they do not unduly limit legitimate use of services, customer satisfaction, and revenue.

In relation to mobile telephone offences, some carriers have refused to allow subscribers to use subscriber identification or 'SIM' cards on other countries' networks without first having undergone special credit checks. In Europe some service providers have even considered the withdrawal of roaming capabilities entirely in order to reduce losses. In the United States, carriers initially dealt with roaming frauds by blocking calls made to those countries most frequently called by offenders, such as Colombia and the Dominican Republic. In France, network operators only permit international use of mobile telephones at the specific request of a subscriber, while in the United Kingdom, Cellnet only allows international use of mobile telephones to specified customers. In order to guard against the payment of accounts with bad cheques, some carriers only allow customers to have network access after cheques have cleared (see Grabosky and Smith 1998). System architecture will provide the most effective solution.

Securities fraud

Crime related to securities markets will elicit a range of responses. Governments will continue to provide online surveillance of markets and investor information in electronic form. State surveillance is already enhanced through the availability of hotlines enabling citizens to report suspicious behaviour promptly. Third-party compliance specialists in security and compliance will be engaged to protect a company's website or other electronic information against illicit interference from internal or external sources, and they can assist in prompt rectification and correction. Multilateral efforts including organizations such as IOSCO will facilitate the detection of and responses to transnational illegality.

As online share trading becomes even more democratized, governments will require those who provide information or access services to day traders to post prominent caveats relating to the risks an unwitting investor faces.

New technologies are also being developed to safeguard against 'web-jacking' (diversion to a substitute site) or counterfeiting of websites. A small software company has developed 'Trustsite Solution', a software package that creates a digital fingerprint for each web page and each piece of content. Another component of the package sets up a validation server that constantly monitors a site's certified content as each page is loaded (Scannell 1999).

Deceptive advertising

Most regulatory agencies throughout the world will provide information in paper form and electronically through websites that alert consumers to misleading and deceptive practices. Consumer groups also represent a good source of trusted information for consumers. Consumer World (2000) has links to over 1400 consumer protection and regulatory agencies.

Technology now gives unprecedented opportunities for disseminating information to consumers, and for enlisting their help in identifying deceptive practices. For example, the National Fraud Information Centre (NFIC) in the United States, which is a project of the National Consumers League, is a non-profit organization that operates a consumer hotline to provide services and help for consumers who may want to file complaints (National Fraud Information Center 2000). The NFIC also sends appropriate information to the fraud database maintained by the FTC and the National Association of Attorneys-General (ACCC 1997).[2]

Many regulatory agencies throughout the world now conduct regular examinations of online content in an attempt to locate illegal and objectionable material. Some, such as the FTC in the United States, have permanent teams of investigators appointed to conduct online surveillance for deceptive and misleading advertising practices (Clausing 1999). The scale of the task is, however, considerable, and action against any objectionable material that is discovered is necessarily limited by jurisdictional boundaries.

In an attempt to widen the scope of online surveillance, a number of regulatory agencies in various countries have undertaken joint surveillance activities, or websurfing days. International Internet Sweep Days are coordinated by the International Marketing Supervision Network, an informal network of consumer law enforcement agencies around the world that cooperate on mutual enforcement matters. Members include the Office of Fair Trading in the United Kingdom, the FTC in the United States and the ACCC in Australia, along with a number of other consumer law enforcement agencies in 29 countries (Office of Fair Trading 1998).

In addition to these coordinated activities, many agencies conduct their own regular surveillance of the Internet. In Australia, for example, the ACCC regularly 'surfs the Net' for suspect sites and has identified hundreds that contain misleading, deceptive and unlawful material. The ACCC website (Australian Competition and Consumer Commission 2000) asks consumers who have encountered a website that they suspect might be illegal to let the

ACCC know by clicking on the 'Slam-a-Scam' icon on the website or by sending the suspect e-mail to sweep.day@accc.gov.au

The most effective strategy, then, is for individuals to be made aware of the risks they face in conducting business electronically and informed about the nature of misleading and deceptive practices which are currently being conducted online. Already there is a lot of such information available. The challenge is to make consumers aware of its existence.

Existing international government organizations such as the OECD Committee on Consumer Policy and International Marketing Supervision Network can also promote international cooperation between government consumer protection and law enforcement agencies. In the future it may be necessary to establish a new, truly international, professional association made up of individual consumer affairs officials to encourage strong networking, a cooperative approach to solving consumer problems and a sharing of ideas (ACCC 1997). Alternatively, it may be necessary to establish international consumer complaints organizations such as the idea of a European Consumer Ombudsman which has been suggested for dealing with cross-border complaints in the European Community (see Goyens and Vos 1991).

Misappropriation of intellectual property

Of all the areas of electronic theft, the misappropriation of intellectual property is perhaps most amenable to private solutions. Here, well-organized copyright ownership interests, from Microsoft to Playboy, alone or reinforced by industry associations (such as the not-for-profit DVD Copy Control Association), are best suited to defending their own interests. They command the resources to detect and investigate violations.

Deterrent effects may also be achieved through the use of technology itself. Technologies are already being developed that enable automated searches for pirated images. One private company had developed software that can identify the source of pirated material; the firm also preserves evidence in the event of a subsequent lawsuit or prosecution (Copyright Control Services 2000).

Another strategy developed to prevent software piracy would entail the use of so-called Logic Bombs, which are installed into programs. When activated through an act of unauthorized copying, the malicious code destroys the copied data and can even damage other software or hardware being used by the offender. Given the potential for collateral damage, the legality of such applications is questionable. Perhaps more appropriate are those technologies which degrade content that is copied without permission.

Industrial espionage

Strategies for the prevention and control of industrial espionage will depend on the legal culture and enforcement capacity of the jurisdiction in question.

Powerful nations such as the United States will threaten the criminal sanction and will assert extraterritorial jurisdiction, as provided by the *Economic Espionage Act* 1996. Other nations will encourage or perhaps even assist their companies in managing risks through enhanced information security technologies. But again, the first line of defence will be self-help. Most large corporations have a large security apparatus; those that do business abroad are often in no position to seek the assistance of law enforcement in foreign countries (who may indeed be engaging in economic espionage themselves). Security of corporate information is, and will remain, fundamentally a private sector responsibility.

Theft of personal information

Misappropriation of personal information (as distinct from illegal interception of communications) would appear most appropriate for resolution by market mechanisms. To be sure, when such misappropriation occurs in furtherance of another criminal matter, such as stalking or harassment, the criminal law can be mobilized, as it can when personal details are copied from corporate databases and sold commercially.

The benchmark in regulatory protection of personal information has been set by the European Union. But regulatory protection is but one element, hardly sufficient, for protecting an individual's personal information. Again, self-help will be important, and individuals should be no less careful with their information in cyberspace than on the ground.

Otherwise, privacy can best be enhanced by a combination of privacy-enhancing technologies and market mechanisms that will reward those who guarantee an individual's privacy and shun those who fail to do so. Just as personal information will become a valuable commodity, to be kept, or traded or sold, so too will privacy protection. Privacy-enhancing technologies will allow people to venture through cyberspace without leaving commercial footprints, and providers will be sought out not only for their product but also for the discretion with which they guard customers' or subscribers' data. These will be reinforced by private certification services, and by industry mechanisms of self-regulation that provide further assurances of privacy protection.

Further Research

Such a rapidly developing subject as electronic crime invites the attention of researchers. As the terrain is new, research possibilities are numerous and diverse. Without wishing to exclude other fruitful avenues of research, we suggest a number of issues that appear to hold out exceptional promise.

First, new organizational forms through which law enforcement agencies interact with non-government actors should be inventoried and catalogued. Just as botanists of today, and their predecessors of years past, enhance our knowledge of the natural world with their efforts to describe and classify, so too should various institutional configurations and patterns of communications, integration, and other forms of interaction be regularly and comprehensively mapped. Above all, these new forms of organizational life should be analysed in terms of their viability and failure, their public accountability and the equity of their impact. One could envisage a continuum from 'pure' self-help on the part of a private entity to all but a state monopoly of electronic crime control. Which organizational form appears best to serve the wider public interest would be worth exploring.

Second, in the current climate of active analysis and reform of existing laws to accommodate the new technologies, the need exists to evaluate the effectiveness of any reforms that have been adopted. Knowing which legislative solutions have been successful and which have failed would not only help those jurisdictions that have yet to amend their laws, but would also enable early remedial legislation to be introduced. Even the relatively simple task of documenting the process of law reform in the digital age should be continued by international research and law reform agencies.

The next area of research would focus on the decision-making criteria used by law enforcement authorities when they consider whether or not to start an investigation. Well over a decade ago, Clifford Stoll (1991) revealed how he encountered great difficulty in attracting the interest of local and national authorities in the United States to a series of computer intrusions in which the value of missing funds was only US$0.75. It became increasingly apparent that the spare change at issue there was only the tip of a much larger iceberg of criminality. How law enforcement agencies manage demands that may exceed their capacity to respond, and how they choose between competing priorities, would make for interesting research. In keeping with our theme of legal pluralism, the extent to which these decisions are based on the presence or absence of complementary external resources would enhance this interest.

Another avenue of research would analyse those cases that are selected by law enforcement agencies for investigation, and would differentiate investigatory success from failure. The fundamental objective of such research would be to predict the outcome of an investigation from certain basic characteristics of a case; in other words, to identify the significant features of a case that will determine success (or otherwise) in investigation and prosecution. Conclusive research findings would provide early indication of likely success or failure of investigations in progress. To the extent that indicators of probable failure are amenable to intervention, findings may help to improve future success rates; to the extent that they are not amenable to intervention, indicators of probable failure may inform resource allocation decisions. Again, the extent to which the availability of pluralistic resources bear on the outcome of an investigation would be worth knowing.

Although only a tiny minority of computer-related crimes result in successful prosecutions, it might be useful to study the sentencing of convicted cybercriminals. Sentences of imprisonment for white-collar offenders are infrequent enough to be remarkable in the English-speaking world. How courts regard electronic theft as opposed to other computer-related crimes, and comparable terrestrial offences, would make for interesting analysis.

The fact that cyberspace knows no geographic frontiers means that a great deal of electronic theft, no less than other forms of computer-related crime, takes place across jurisdictions. While it may be premature to study dispositional alternatives, precisely how the law is mobilized against transnational offenders, and what mechanisms may be emerging to replace traditional forms of mutual assistance and extradition may provide a roadmap for the future.

Conclusion

Prediction is always risky, and predicting the future of electronic theft is likely to be especially difficult. The pace of technological change seems unlikely to abate. The digital environment seems likely to be characterized by increasing integration and concentration of information and mass media industries. At the same time, new products will emerge from modest beginnings; not that long ago, both Microsoft and Apple Computer were 'backyard' enterprises. Their successors may well produce technologies that are no less revolutionary.

Just as regulation by law alone is not feasible, so too is regulation by technology and by market forces. As Lessig (1999) observes, we are experiencing a shift in regulatory power. We join Lessig in acknowledging that the leading role will in future be played by technology, reinforced by the rule of law. As Streeck and Schmitter (1985: 25) observe, 'A state that withdraws, in selected areas, from direct to procedural control does not become a weak state; in terms of the effectiveness of its policies, it may in fact gain in strength'. As Foucault would observe, maximizing the autonomy or self-perceived autonomy of regulatory targets becomes a key term in the exercise of political power. One might speak of governing people by their freedom to choose (Rose and Miller 1992: 174).

The question of whether public needs should be met by government dictates on the one hand, or by markets on the other, has become a misplaced question. We must now inquire what institutional form, or, even more appropriately, what blend of institutional forms, is best suited to a given task. The design and guidance of hybrid regulatory systems will be an essential activity of government in the years ahead.

To reduce the risk of electronic theft in the twenty-first century, it will be essential for all those involved to work cooperatively in making use of the latest technologies of information security. Technology has not only provided a new medium in which property crime may take place but has also enhanced our ability to detect and to control it. The use of reliable and current information will, in the end, be the most effective means of preventing some of the more egregious frauds that have taken place in the past.

The most effective strategy, then, is for individuals and companies to be made aware of the risks they face in conducting business electronically and to be informed about the nature of unlawful practices that are currently being conducted online.

The transnational nature of much electronic theft, a theme that has been evident through most of this book, is fostering international harmonization in many areas of law enforcement. The movement is gradual, but will lead to a degree of consistency, if not uniformity. Efforts to arrange broadly comparable legislation relating to the substantive criminal law, as well as the law of cross-border search and seizure, will make the detection, investigation, and prosecution of electronic thieves somewhat easier in future. A greater degree of harmonization will almost certainly occur in information security, as markets deliver solutions where laws cannot. The result will be a world in which honest individuals and organizations will be better equipped to protect themselves.

Notes

1 The cause of action whereby citizens were empowered to sue on behalf of the state became known as Qui Tam (qui tam pro domino rege quam pro se ipso in hac parte sequitur) which may be translated as 'who brings the action for the king as well as for oneself' (Caminker, 1989). See also Australian Law Reform Commission (1985).

2 See also the FBI's Fraud Complaint Center at http://www.ifccfbi.gov/ (visited 7 December 2000).

References

Abbreviations
AGPS Australian Government Publishing Service
HMSO Her Majesty's Stationery Office

ACCC (Australian Competition and Consumer Commission) 1997a, *Advertising and Selling*, ACCC, Sydney.

ACCC (Australian Competition and Consumer Commission) 1997b, *The Global Enforcement Challenge: The Enforcement of Consumer Protection Laws in a Global Marketplace – Discussion Paper*, ACCC, Sydney.

ACCC (Australian Competition and Consumer Commission) 1998, *Annual Report 1997–98*, AusInfo, Canberra.

ACCC (Australian Competition and Consumer Commission) 1999, *Internet Service Providers*. http://www.accc.gov.au/docs/catalog.htm (visited: 30 April 1999).

ACCC (Australian Competition and Consumer Commission) 2000 *International Internet Sweep Day*. http://www.accc.gov.au/ecomm/access.htm (visited 24 July 2000)

ACT Government 2000, Kiosk Locations. http://www.act.gov.au/austouch/austouch.html (visited 25 October 2000).

AICPA (American Institute of Certified Professional Accountants) 2000 *CPA Webtrust*. http://www.cpawebtrust.org/ (visited 24 October 2000)

Alexander, M. 1995, *The Underground Guide to Computer Security*, Addison-Wesley Longman Inc., New York.

America OnLine 2000 *Privacy Policy*. http://www.aol.com/info/privacy.html (visited 24 October 2000).

American Society for Industrial Security, International 1998, *Trends in Intellectual Property Loss Survey*. http://www.asisonline.org (visited 17 December 1999).

Anon. 1996, 'Fingerscan's $2.5m Deal', *Security Australia* 16(10): 2.

Anon. 1997, 'Feds Expose Internet Ring that Cost Thousands in Phone Bills', *Detroit News*, 20 February.

Anon. 1999, 'Minnesota A-G Sues U.S. Bank for Selling Out Customers to Telemarketers', *Corporate Crime Reporter* 14(24): 1.

Arlidge, A., J. Parry and I. Gatt 1996, *Arlidge and Parry on Fraud*, 2nd edn, Sweet & Maxwell, London.

Arrow, K. J. 1962, 'Economic Welfare and the Allocation of Resources for Invention'. In *The Rate and Direction of Inventive Activity: economic and social factors*. A conference

of the Universities–National Bureau Committee for Economic Growth of the Social Science Research Council. Princeton University Press, Princeton, p. 609.
Associated Press 1998a, 'Computer crook turned informant on the lam again', *CNN Interactive*, 1 December.
Associated Press 1998b 'GeoCities settles Internet privacy case with regulators', 13 August. http://archives.seattletimes.nwsource.com/news/technology/html98/geo081398.html (visited 24 October 1999).
Associated Press 1999a, 'Pyramid Schemes in Cyberspace: The Same Old Deal'. http:// www.jamcaster.com/TechNews9903/11_pyramid.html (visited 24 October 2000) .
Associated Press 1999b, 'Internet Hoax Sends Technology Stock Soaring', Associated Press, 8 April. http://www. acmi.canoe.ca/TechNews9904/08_hoax.html (visited 24 October 2000).
Associated Press 1999c, 'Internet Hoaxer Sentenced', Associated Press, 30 August. http://www.usatoday.com/life/cyber/tech/ctf985.html (visited 24 October 1999).
Associated Press 1999d, Microsoft, Web's Biggest Advertiser, to Require Privacy Promises (23 June). http://www.nytimes.com/library/tech/99/06/biztech/articles/23soft-privacy.html (visited 24 October 1999).
Associated Press 1999e, 'Man guilty of Internet stalking'. http://www.bayinsider.com/news/1999/01/20/stalking.html (visited 1 July 1999).
Audio and Video 2000 *Reference*: MPEG. http://www.englishmedia.com/av/reference.htm (visited 24 October 2000).
Audit Commission 1994, *Opportunity Makes a Thief: An Analysis of Computer Abuse*, HMSO, London.
.au Domain Administration (2000), '.au Domain Administration Home Page'. http://www.auda.org.au/ (visited 24 October 2000).
Australia, Attorney-General's Department 1998, *Electronic Commerce: Building the Legal Framework*, Report of the Electronic Commerce Expert Group to the Attorney-General, AGPS, Canberra. http://www.law.gov.au/aghome/advisory/eceg/ecegreport.html (visited 24 October 2000).
Australia, Model Criminal Code Officers Committee, Standing Committee of Attorneys-General 1994, *Model Criminal Code, Chapter 3: Blackmail, Forgery, Bribery and Secret Commissions: Discussion Paper*. Attorney-General's Department, Canberra.
Australia, Model Criminal Code Officers Committee 1995, *Model Criminal Code Chapter 3: Theft, Fraud, Bribery and Related Offences*, Final Report, AGPS, Canberra.
Australia, Model Criminal Code Officers Committee 2000, *Damage and Computer Offences: Discussion Paper, Chapter 4*, Attorney-General's Department, Canberra.
Australia, Office of Government Information Technology 1998, *Gatekeeper: A Strategy for Public Key Technology Use in the Government*, AGPS, Canberra.
Australia, Office of Strategic Crime Assessments and Victoria Police Computer Crime Investigation Squad 1997, Computer Crime and Security Survey, Attorney-General's Department, Canberra.
Australia, Privacy Commissioner 2000, *The Australian Privacy Commissioner's Website*. http://www.privacy.gov.au (visited 24 October 2000).
Australian Bureau of Statistics 2000, *Use of the Internet by Householders, Australia, November 1999* (Cat. No. 8147.0), Australian Bureau of Statistics, Canberra.
Australian Customs Service 1998, *Annual Report 1997–98*, Australian Customs Service, Canberra.
Australian Domain Name Administration 2000. http://www.auda.org.au/archive/adna/ (visited 24 October 2000).
Australian Federal Police 1996, *Annual Report 1995–96*, AGPS, Canberra.
Australian Federal Police 1997, *Annual Report 1996–97*, AGPS, Canberra.
Australian Federal Police 1998, *Annual Report 1997–98*, AGPS, Canberra.

Australian National Audit Office 1994, *The Australian Government Credit Card: Some Aspects of its Use*, Audit Report No. 1, 1993–94, Project Audit, AGPS, Canberra.

Australian Payments System Council 1998, Annual Report 1997–98, Reserve Bank of Australia, Sydney.

Australian Securities Commission 1999, *Policy Statement 118, 'Investment Advisory Services: Media Computer Software and Internet Advice'*, ASC Digest PS 7/1761.

AUSTRAC (Australian Transaction Reports and Analysis Centre) 1999, 'Great Tax Results', *AUSTRAC Newsletter*, Spring, p. 1.

Ayres, I. and J. Braithwaite 1992, *Responsive Regulation: Transcending the Deregulation Debate.* New York: Oxford University Press.

Baer, P. 1996, 'The Australian Federal Police and Commonwealth Department Security Management', *Platypus Magazine (Journal of the Australian Federal Police)*, 50 (March): 22–6.

Bainbridge, David, and Graham Pearce 1998, 'Data Protection: The UK Data Protection Act 1998 – Data Subjects' Rights', *Computer Law and Security Report* 14(6): 401–6.

Banisar, David 1999, 'Privacy and Data Protection Around the World'. Paper presented to 21st International Conference on Privacy and Personal Data Protection, Hong Kong, 13 September. http://www.pco.org.hk/conproceed.html (visited 24 October 2000).

Banner, S. 1998a, 'The Origin of the New York Stock Exchange, 1791–1860', *Journal of Legal Studies* 27: 113–40.

Banner, S. 1998b, *Anglo-American Securities Regulation: Cultural and Political Roots, 1690–1860*, Cambridge University Press, Cambridge.

Bardach, E., and R. Kagan 1982, *Going by the Book: The Problem of Regulatory Unreasonables*, Temple University Press, Philadelphia.

Barker, G. 2000, 'NZ Traders Reach for the Mobile', *The Age* (Melbourne) 22 March, p. 5.

Barker, G. and L. Johnson 1999, 'Private Profiles Laid Bare in Online Treasure Troves', *The Age* (Melbourne), 1 December, p. 2.

Barlow, J. P. 1994, 'The Economy of Ideas: A framework for patents and copyrights in the Digital Age', *Wired* 2.03 (March 1994), at http://www.wired.com/wired/archive/2.03/economy.ideas_pr.html.

Barnett, Amanda 1999, 'Hundreds of defendants named in lawsuit over DVD hacking', CNN December 28, 1999. http://cgi.cnn.com/1999/TECH/ptech/12/28/dvd.crack/index.html (visited 24 October 2000)

Barry, Scott 1997, 'Spinning a Web of Retaliation', *Seattle Times* 20 February 1997. http://archives.seattletimes.com/extra/browse/html97/shee_022097.html (visited 24 October 2000).

Bauer, James 1998, Testimony to the Subcommittee on Technology, Terrorism and Government Information, Committee on the Judiciary, United States Senate, 20 May, 1998. http://www.securitymanagement.com/library/bauer.html (visited 24 October 2000).

Bellcore 1996, 'New Crypto-Attack Weakens Seeming Strength in Smart Cards, Secure ID Cards, or Vale Cards'. http://www.info-sec.com/crypto/infosec4.html-ssi (visited 24 October 2000).

Bennahum, D. 1999, 'Daemon Seed: Old Email Never Dies', *Wired*, 7.05 (May) 100–11.

Bernstein, Nina 1996, 'On Prison Computer, Files to Make Parents Shiver', *New York Times* 18 November, A1.

Bernstein, Nina 1997, 'Lives on File: Privacy Devalued in Information Economy', *New York Times* June 12. http://www.urbsoc.org/courses/internet/sources/privacy.html (visited 24 October 2000).

Birks, P. 1979, *An Introduction to the Law of Restitution*, Oxford University Press, Oxford.

Black, Donald 1984, 'Social Control as a Dependent Variable'. In Donald Black (ed.), *Toward a General Theory of Social Control*, vol. 1, Academic Press, Orlando, pp. 1–36.

Bowes, C, 1996, 'Digital Dollars', *The Bulletin*, 20 August, p. 50.

Boyer, B., and E. Meidinger 1985, 'Privatizing Regulatory Enforcement', *Buffalo Law Review* 34: 833–964.

Braithwaite, J., and P. Drahos 2000, *Global Business Regulation*, Cambridge University Press, Cambridge.

Brandt, A. 1975, 'Embezzler's Guide to the Computer', *Harvard Business Review* 53: 79–89.

Bridgeman, J. S. 1997, 'Keynote Speech to the Electronic Shopping Forum', *Fair Trading Magazine*, 6 May, Office of Fair Trading, London.

Bridges, M. J., and P. Green 1998, 'Tax Evasion and the Internet', *Journal of Money Laundering Control* 2(2): 105–14.

Burchell, Graham 1991, 'Peculiar Interests: Civil Society and Governing "the System of Natural Liberty"'. In Burchell et al., *The Foucault Effect*, pp. 119–150.

Burchell, Graham, Colin Gordon and Peter Miller (eds), *The Foucault Effect: Studies in Governmentality*, Harvester Wheatsheaf, London.

Butler, A. 1996, 'Regulation of Content of On-line Information Services: Can technology itself solve the problem it has created?', *University of New South Wales Law Journal* 19(2): 193–221.

Byford, Katrin 1998, 'Privacy in Cyberspace: Constructing a Model of Privacy for the Electronic Communications Environment', *Rutgers Computer and Technology Law Journal* 24: 1–74.

Canadian Security Intelligence Service 1998, Transnational Criminal Activity (November 98). http://www.csis-scrs.gc.ca/eng/backgrnd/back10e.html (visited 24 October 2000).

Campbell, Duncan 1999, 'Careful, they might hear you', *The Age* (Melbourne) 23 May 1999. http://www.theage.com.au/daily/990523/news/news3.html (visited 17 October 2000).

Campbell, Duncan 2000, 'Big Brother is Back', *Sydney Morning Herald*, 29 August, Computer Section, p. 1.

Campbell, R. 1999, 'DOFA Review in Wake of Alleged $8m Fraud', *Canberra Times* 17 February, pp. 1–2.

Carter, S. 1996, 'Online "Bank" Cashes in on Cyber Commerce', *Australian*, 30 July, Computers, p. 49.

Cate, Fred H. 1998, 'Privacy and Telecommunications', *Wake Forest Law Review*, 33(1): 1–49.

CAUDIT (Committee of Australian University Directors of Information Technology) 1997, Code of Practice Relating to Content That May Infringe Censorship Laws, 3 April. http://www.caudit.edu.au/caudit/codes/index.html (visited 2 November 2000).

Cavoukian, A. 1999, 'Privacy and Biometrics'. Paper presented to 21st International Conference on Privacy and Personal Data Protection, Hong Kong, 13 September. http://www.pco.org.hk/conproceed.html (visited 24 October 2000).

Cella, J. J., and J. R. Stark 1997, 'SEC Enforcement and the Internet: Meeting the challenge of the next millennium', *The Business Lawyer* 52: 815–49.

Centrelink 1997, *Data-Matching Program: Report on Progress 1996–97*, Data-Matching Agency, Canberra.

Chapman, M. 2000, 'Can a Computer be Deceived? Dishonesty offences and the electronic transfer of funds', *Journal of Criminal Law* 64(1): 89–97.

Chin, Ko-Lin 1996, *Chinatown Gangs: Extortion, Enterprise and Ethnicity*, Oxford University Press, New York.

Clarke, Roger 1998a, 'Direct Marketing and Privacy'. Paper presented at AIC Conference 'The Direct Distribution of Financial Services', Sydney, 24 February 1998. http://www.anu.edu.au/people/Roger.Clarke/DV/DirectMkting.html (visited 24 October 2000).

Clarke, Roger 1988b, Information Technology and Dataveillance. Commun. ACM 31,5 (May 1988) 498–512. http://www.anu.edu.au/people/Roger.Clarke/DV/CACM88.html (visited 24 October 2000).

Clarke, Roger 1998c, Cookies. http://www.anu.edu.au/people/Roger.Clarke/II/Cookies.html (visited 24 October 2000).

Clarke, Roger, and Gillian Dempsey 1998, 'Technological Aspects of Internet Crime Prevention', Australian Institute for Criminology Conference on Internet Crime, Melbourne, 16–17 February.

Clarke, Roger, and Gillian Dempsey 1999, 'Electronic Trading in Copyright Objects and Its Implications for Universities'. http://www.anu.edu.au/people/Roger.Clarke/EC/ETCU.html (visited 24 October 2000).

Clarke, Ronald V. 1995, 'Situational Crime Prevention'. In Michael Tonry and David Farrington (eds), *Building a Safer Society: Strategic Approaches to Crime Prevention.* University of Chicago Press, Chicago, pp. 91–150.

Clausing, J., 1999a, 'FTC Appoints Team to Monitor Advertising Online', *New York Times,* 11 May.

Clausing, J., 1999b, 'FTC Holds Meeting on International E-Commerce', *New York Times,* 8 June.

Clough, B., and P. Mungo 1992, *Approaching Zero: Data Crime and the Computer Underworld,* Faber & Faber, London.

CNN (Cable News Network) 1999, 'Business manager linked to prostitute through Hotmail hole', CNN Online, September 3, 1999 Web posted at: 12:49 p.m. EDT (1649 GMT). http://www.cnn.com/TECH/computing/9909/03/hotmail.fallout/ (visited 24 October 2000).

Coffee, J. C. 1997, 'Brave New World? The impact(s) of the internet on modern securities regulation', *The Business Lawyer* 52(August): 1195–233.

Cohen, L., and M. Felson 1979, 'Social change and crime rate trends: A routine activity approach', *American Sociological Review,* 44: 588–608.

Commodity Futures Trading Commission 2000, *Enforcement Program.* http://www.cftc.gov/enf (visited 24 October 2000).

Consumer World 2000, Consumer Agencies and Organizations http://www.consumerworld.org/pages/agencies.htm (visited 24 October 2000).

Cook, V. 1999, 'Trust Me, I'm a Computer', *Communications Newsletter* September: 14–15.

Cope, N. 1996, *Retail in the Digital Age,* Bowerdean Publishing Co. Ltd, London.

Copyright Control Services 2000, 'Protecting Copyright Online'. http://www.copyrightcontrol.com/ (visited 24 October 2000).

Council of Europe 1995, *Problems of Criminal Procedural Law Connected with Information Technology,* Recommendation No. R (95) 13, Council of Europe Publishing, Strasbourg.

Council of Europe 2000 Draft Convention on Cyber-Crime (Draft no. 22 REV) http://conventions.coe.int/treaty/EN/cadreprojets.htm (visited 26 October 2000).

Cybank 1997 *Welcome to Cybank.* http://www.cybank.net (visited 23 July 2000).

Daly, K. 1989, 'Gender and Varieties of White-Collar Crime', *Criminology* 27: 769–93.

Dancer, Helen 1999, 'Can we trust in TRUSTe?', *Bulletin,* 11 May, p. 85.

Da Silva, W. 1996, '"Hackers" May Evade Charges', *The Age* (Melbourne) 11 June, p. C1.

Da Silva, W. 1996, 'Con Artists of the Internet', *The Age* (Melbourne) 10 December, pp. D1 and 7.

Davies, L. 1997, 'Contract Formation on the Internet: Shattering a Few Myths'. In L. Edwards and C. Waelde (eds), *Law and the Internet: Regulating Cyberspace,* Hart Publishing, Oxford, pp. 97–120.

Davies, Simon G. 1997, 'Re-Engineering the Right to Privacy: How Privacy has been Transformed from a Right to a Commodity'. In Philip Agre and Marc Rotenberg (eds), *Technology and Privacy: the New Landscape*, MIT Press, Cambridge, Mass., pp. 143–67.

Davis, Patricia 1998, 'Peeping Toms with Videocams Plague Malls', *Seattle Times*, 9 June. http://seattletimes.nwsource.com/news/nation-world/html98/peep_060998.html (visited 24 October 2000).

DCTIA (Australia, Department of Communications, Information Technology and the Arts) 2000a, *Website seals of approval: a comparative examination.* http://www.dcita.gov.au/nsapi-graphics/?MIval=dca_dispdoc&pathid=%2fshoponline%2fsealtable%2ehtml (visited 24 October 2000).

DCTIA (Australia, Department of Communications, Information Technology and the Arts) 2000b, *Shopping on the Internet: Facts for Consumers.* http://www.dcita.gov.au/cgi-bin/trap.pl?path=3815 (visited 24 October 2000).

Deakin University 1994, *Fraud Against Organisations in Victoria*, Deakin University, Geelong.

Delahaye 2000, Delahaye Medialink. http://www.delahaye.com (visited 24 October 2000).

Denning, D. E. 1998, 'Cyberspace Attacks and Countermeasures'. In D. E. Denning and P. J. Denning, *Internet Besieged: Countering Cyberspace Scofflaws*, ACM Press, New York, pp. 29–55.

Denning, D. E. 1999, *Information Warfare and Security*, Addison Wesley, Boston.

Demsetz, H. 1969, 'Information and Efficiency: Another Viewpoint', *Journal of Law and Economics* 12: 1.

Diffie, Whitfield, and Susan Landau 1998, *Privacy on the Line: The Politics of Wiretapping and Encryption.* MIT Press, Cambridge, Mass.

Digicash 2000, *eCash: Electronic Payment Solutions.* http://www.digicash.com/ (visited 23 July 2000).

Digimarc 2000, *Getting a Watermarking Subscription and a Digimarc ID.* http://www.digimarc.com/imaging/pridsignup.shtml (visited 24 October 2000).

Duncan, M. 1998, 'Collaborating to Catch Currency Counterfeiters', *RCMP Gazette* 60(5–6): 52–4.

Duric, Z., N. F. Johnson and S. Jajodia 1999, Recovering Watermarks from Images (unpublished manuscript).

Duva, J. 1997, 'Online Hacker Pleads Guilty to Felony Computer Fraud' 1997 B.C. *Intellectual Property and Technology Forum 012301.* http://www.bc.edu/bc_org/avp/law/st_org/iptf/headlines/content/1997012301.html (visited 24 October 2000).

Dyson, Esther 1998, *Release 2.1: A Design for Living in the Digital Age*, Broadway Books, New York.

Eaton, L. 1996, 'Let the Cyberinvestor Beware: A tale of stock promotion, regulation and the internet', *New York Times*, 5 December, p. D1.

EBay 2000, *e-Bay: The World's Online Marketplace.* http://pages.ebay.com/index.html (visited 24 October 2000).

Eden, S. 1997, 'The Taxation of Electronic Commerce'. In L. Edwards and C. Waelde, *Law and the Internet: Regulating Cyberspace*, Hart Publishing, Oxford, pp. 151–79.

Ehrlich E. 1912, *Fundamental Principles of the Sociology of Law*, Harvard University Press, Cambridge, Mass.

Ellsberg, Daniel 1968, *The Theory and Practice of Blackmail.* Rand Paper P-3883, RAND Corporation, Santa Monica.

Endfraud.com 2000, *Endfraud.com.* http://www.endfraud.com/ (visited 24 October 2000).

ERights.org 2000, *E: Open Sourced Distributed Capabilities.* http://www erights.org (visited 23 July 2000).

Errico, Marcus 1997, 'Elle Macpherson Victim of Extortion Plot', E! Online, 8 July. http://ace1-eol1.eonline.com/News/Items/0,1,1410,00.html (visited 24 October 2000).

Espionage Unlimited 2000, *Surveillance Equipment.* http://www.espionage-store.com/surveillance.html (visited 23 July 2000).

eWatch 2000, *eWatch* http://www.ewatch.com/ (visited 24 October 2000).

Fact Net 1996, 'The Shetland Copyright Battle' (6 December). http://www.factnet.org/shetlnd2.html (visited 24 October 2000).

FBI (Federal Bureau of Investigation) 2000, *Carnivore Diagnostic Tool.* http://www.fbi.gov/programs/carnivore/carnivore2.htm (visited 2 November 2000).

Fenwick and West 1998, *Securing and Protecting a Domain Name for Your Website.* http://www.fenwick.com/pub/domain.htm (visited 12 December 2000).

Festa, P. 1997, 'Short Take: AOL Hacker Sentenced', *CNET News.* http://news.cnet.com/news/0-1005-200-317701.html (visited 24 October 2000).

Fields, B. 1997 'Phishing for Passwords on AOL: How Hackers Wreak Havoc' *PC World Online* 27 June. http://www.pcworld.com/news/daily/data/0697/970627185134.html (visited 23 July 2000).

Finn, P. D. 1979 *Fiduciary Obligations,* Law Book Company, Sydney.

Finn, P. D. 1984, 'Confidentiality and the "Public Interest"', *Australian Law Journal* 58: 497.

Fisse, B. 1990, *Howard's Criminal Law,* 5th edn, Law Book Co. Ltd, Sydney.

Fitzpatrick, P 1984, 'Law and Societies', *Osgoode Hall Law Journal* 22: 115–38.

Flaherty, David 1989, *Protecting Privacy in Surveillance Societies,* University of North Carolina Press, Chapel Hill.

Fontana, H. 1998, 'Securities on the Internet: World Wide Opportunity or Web of Deceit?', *Inter-American Law Review* 29(1–2): 297–328.

Forbes, M. 1999, 'Fraud Check on Kennett Staff', *The Age* (Melbourne) 30 March, pp. 1–2.

Foster, Bradley 1998, 'UW student charged in extortion attempt', *University of Washington Daily,* 19 August. http://www.thedaily.washington.edu/archives/1998_summer/Aug.19.98/soda.html (visited 24 October 2000).

Foucault, Michel 1977, *Discipline and Punish: The Birth of the Prison.* Pantheon, New York.

Foucault, Michel 1979, 'Governmentality'. In Burchell et al., *The Foucault Effect,* pp. 87–104.

Foucault, Michel 1980 *Power/Knowledge: Selected Interviews and Other Writings* (ed. Colin Gordon) Harvester Press, Brighton.

419 Coalition 2000, *Nigeria: The 419 Coalition Website.* http://home.rica.net/alphae/419coal/Extortionsamples.htm (visited 24 October 2000).

FSA (UK Financial Services Authority) 1998, 'Treatment of Material on Overseas Internet World Wide Web Sites Accessible in the UK but not Intended for Investors in the UK', Guidance Release 2/98, May.

FSA (UK Financial Services Authority) 2000, http://www.fsa.gov.uk/consumer/whats_new/index.html (visited 24 October 2000).

FTC (US Federal Trade Commission) 1999, http://www.ftc.gov/ (visited 24 October 1999).

FTC (US Federal Trade Commission) 2000, *Day Trading Ads: Cutting Through the Bull.* www.ftc.gov/bcp/conline/edcams/daytrade (visited 24 October 2000).

Galanter, Marc, and David Luban 1993, 'Poetic Justice: Punitive Damages and Legal Pluralism', *American University Law Review* 42: 1393–463.

Gambetta, Diego 1993, *The Sicilian Mafia: The Business of Private Protection,* Harvard University Press, Cambridge, Mass.

Garfinkel, S. 1998, 'Read them and weep', *Boston Globe* 7 May 1998. http://simson.net/chips/98.Globe.05-07.Read them and weep.htm (visited 7 December 2000).

Garland, David 1996, 'The Limits of the Sovereign State: Strategies of Crime Control in Contemporary Society', *British Journal of Criminology* 36(4): 445–71.

Garrow, David J. 1999, *The FBI and Martin Luther King, Jr.,* W. W. Norton & Co., New York.

Gavison, Ruth 1980, 'Privacy and the Limits of Law', *Yale Law Journal* 89(3): 421–71.

Gengler, B. 1999, 'High-Tech Crime Costs $5bn a Year', *Australian* (Computers), 23 March, p. 9.

Gindin, Susan E. 1997, 'Lost and Found in Cyberspace: Informational Privacy in the Age of the Internet', *San Diego Law Review* 34: 1153–223.

Girard, Kim 1999, 'IBM to Pull Web Ads Over Privacy concerns'. http://news.com/News/Item/0,4,34470,00.html?st.ne.ni.rel (visited 24 October 2000).

Glave, J. 1998, 'Cracker Fights for Flat Rates', Wired News, 20 August. http://www.wired.com/news/technology/0,1282,14556,00.html (visited 24 October 2000).

Gleick, J. 1998, 'Fast Forward: Meaning-Free Capital', *New York Times Magazine*, 7 June.

Glod, Maria 1999, 'Spouses may delete their marriage, but e-mail lives on as evidence', *Seattle Times* 28 April. http://seattletimes.nwsource.com/news/nation-world/html98/mail_19990428.html (visited 24 October 2000).

Goldring, J., L. W. Maher, J. McKeough and G. Pearson 1998, *Consumer Protection Law*, 5th edn, Federation Press, Sydney.

Goldwasser, V. 1998, 'The Regulation of Stock Market Manipulation and Short Selling in Australia'. In G. Walker, B. Fisse and I. Ramsay (eds), *Securities Regulation in Australia and New Zealand*, 2nd edn, LBC Information Services, Sydney, pp. 515–55.

Goyens, M., and E. Vos 1991, 'Crossborder Consumer Complaints Handling in the European Economic Community: The Current Factual Situation and Some Suggestions for Improvement', *European Consumer Law Journal* 4: 193–204.

Grabosky, P., and R. G. Smith 1997, 'Telecommunications and Crime: Regulatory Dilemmas', *Law and Policy* 19(3): 317–41.

Grabosky, P., and R. G. Smith 1998, *Crime in the Digital Age: Controlling Telecommunications and Cyberspace Illegalities*, Federation Press, Sydney.

Grabosky, P., and A. Sutton (eds) 1989, *Stains on a White Collar: Fourteen Studies in Corporate Crime or Corporate Harm*, Federation Press, Sydney.

Grabosky, P., and P. Wilson 1989, *Journalism and Justice: How Crime is Reported*, Pluto Press, Sydney.

Grainger, G. 1999, 'A Co-regulatory Scheme for Internet Content: The Australian Approach'. Paper presented to the Internet Content Summit 1999: Policy Panel, Munich, 9–11 September. http://www.aba.gov.au/about/public_relations/speeches (visited 20 January 2000).

Great Britain 1990, Report of the Committee on Privacy and Related Matters (Chairman David Calcutt QC). Cmnd. 1102, HMSO, London.

Grossman, Wendy 1997, *Net.Wars*, New York University Press, New York.

Gup, B. E. 1995, *Targeting Fraud: Uncovering and Deterring Fraud in Financial Institutions*, Probus Publishing Co., Chicago.

Gurry, D. 1984, *Breach of Confidence*, Oxford University Press, Oxford.

Hafner, K., and J. Markoff 1991, *Cyberpunk: Outlaws and Hackers on the Electronic Frontier*, Simon & Schuster, New York.

Hagel, John III, and Jeffrey Rayport 1997, 'The Coming Battle for Customer Information', *Harvard Business Review* January–February: 53–65.

Haines, J., and P. Johnstone 1998, 'Global Cybercrime: New Toys for the Money Launderers', *Journal of Money Laundering Control* 2(4): 317–25.

Haines, Thomas 1997, 'Suspect charged in threats on Gates', *Seattle Times*, 16 May.

Hall, R. 1993, 'Sydney's Original Sins', *The Age* (Melbourne) *Good Weekend*, 20 November, pp. 72–5.

Hanlon, M. 1999, 'The Touch-Feely Mouse', *The Express* (London), 8 July, p. 22.

Hansell, S. 1996, 'AT&T and Wells Fargo Investing in an Electronic Cash Card', *New York Times*, 19 July, p. C2.

Harlow, C., and R. Rawlings 1992, *Pressure Through Law*, Routledge, London.

Harris, J. 1998, *Industrial Espionage and Technology Transfer: Britain and France in the Eighteenth Century*, Ashgate, Aldershot.

Hart, Michael 1998, Corporate liability for employee use of the Internet and e-mail: steps to take to reduce the risks', *Computer Law and Security Report* 14(4): 223–31.

Harvard Law Review 1996, 'Recent Agency Action: Securities Law: SEC Allows Internet-Based Trading of Securities', *Harvard Law Review* 110: 959–64.

Hauch, Jeanne M. 1994, 'Protecting Private Facts in France: The Warren and Brandeis Tort is Alive and Well and Flourishing in Paris', *Tulane Law Review* 68: 1219–301.

Helft, Miguel 1997, 'Netscape Dodges Bug, "Extortion" Bullet', 13 June (Wired News 2 February 1999).

Hepworth, Mike 1975, *Blackmail: Publicity and Secrecy in Everyday Life*, Routledge & Kegan Paul, London.

Hesseldahl, A. 1998, 'Kicking Crackers Off the Grid', Wired News, 5 August. http://www.wired.com/news/politics/0,1283,14197,00.html (visited 24 October 2000).

Heywood, J. 1999, 'Yahoo! First with Auctions', *The Age* (Melbourne) 11 May.

HIC (Australian Health Insurance Commission) 1997, *Annual Report 1996–97*, AGPS, Canberra.

HIC (Australian Health Insurance Commission) 1998, *Annual Report 1997–98*, AGPS, Canberra.

Higgs, P. 1999, Issues for Australian Competitiveness in the Knowledge Based Economy (unpublished working paper, IPR Systems).

Highland, H. J. 1997, 'The Threats on the Web', *Computer Fraud and Security* June: 7–10.

Hildreth, Peter 1999, Testimony before the Permanent Subcommittee on Investigations, Governmental Affaits Commiteee, US Senate, March 23. http://www.nasaa.org/whoweare/speeches/PSI.html (visited 18 December 1999).

Hipkin, S. 1999, 'Identifying and Preparing to Combat the Next Stages of Fraud Migration: The drift towards "process fraud"', Vodaphone Targeting Fraud ICM Conference, Melbourne, 15–16 September.

Hirschliefer, J. 1971, 'The Private and Social Value of Information and the Reward to Inventive Activity', *American Economic Review* 61: 561.

Hirschliefer, J. 1973, 'Where Are We Now in the Theory of Information?', *American Economic Review* 63: 31.

Hogan, M. 1997, 'Cyber Scams', *PC World*, May: 169–80.

Holland, K. 1995, 'Bank Fraud, The Old-Fashioned Way', *Business Week* 4 September, p. 88.

Hook, Patrick 1998, 'The Euro: A Crooks' Charter?', *Police* 30(11): 6–7.

Hughes, G. 1998, 'Tax Staff Probed for Fraud', *The Age* (Melbourne) 16 September, p. 1f.

Hyatt, J. 1993, 'NASD Disciplines Individuals, Firms Charged with Securities Violations', *Wall Street Journal* 20 December, p. A7A.

IBM 1996, *IBM and Xerox Intend to Join Forces for Intellectual Property Protection on the Internet.* http://www.ibm.com/news/iworld/501xer.html (visited 24 October 2000).

ICAC (NSW Independent Commission Against Corruption) 1992, *Report on Unauthorised Release of Government Information* (Roden Report), ICAC, Sydney.

International Marketing Supervision Network 2000, *International Marketing Supervision Network.* http://www.imsnricc.org/ (visited 24 October 2000).

Internet Industry Association 1999, *Code of Practice.* http://www.iia.net.au/index2.html (visited 24 October 2000).

IOSCO (International Organization of Securities Commissions) 1997, *Report on Enforcement Issues Raised by the Increasing Use of Electronic Networks in the Securities and Futures Field*, Report of the Technical Committee, September. http://www.iosco.org/docs-public/1997-report_on_enforcement_issues.html (visited 24 October 2000).

ITAA (Information Technology Association of America) 1998, 'Digital © Protection: Intellectual Property Protection In Cyberspace: Towards a New Consensus', Discussion Paper. http://www.itaa.org/copyrite.htm (visited 24 October 2000).

Jackson, Margaret 1997, 'Data Protection Regulation in Australia after 1988', *International Journal of Law and Information Technology* 5(2): 158–91.

James, M. 1995, 'Preventing the Counterfeiting of Australian Currency', in P. Grabosky and M. James (eds), *The Promise of Crime Prevention: Leading Crime Prevention Programs*, Australian Institute of Criminology, Canberra, pp. 12–13.

Jenkins, P. 1999, 'VAT and Electronic Commerce: The Challenges and Opportunities', *VAT Monitor* 10(1): 3–5.

Jesilow, P., H. N. Pontell and G. Geis 1992, *Prescriptions for Profit: How Doctors Defraud Medicaid*, University of California Press, Berkeley.

Johnson, D. 1997, 'More Counterfeiting Cases Involve Computers', *New York Times* 18 August.

Johnson, E. 1996, 'Body of Evidence: How Biometric Technology Could Help in the Fight Against Crime', *Crime Prevention News* December: 17–19.

Johnson, N. 2000, *Steganography*. http://www.jjtc.com/stegdoc/stegdoc.html (visited 24 October 2000).

Johnstone, P. 1999, 'Serious White Collar Fraud: Historical and Contemporary Perspectives', *Crime, Law and Social Change* 30(2): 107–30.

Jones, G., 1970, 'Restitution of Benefits Obtained in Breach of Another's Confidence', *Law Quarterly Review* 86: 463.

Joyce, A. 1999, 'Cautionary Tales of Commonwealth Credit Card Fraud', *Comfraud Bulletin* 12(January): 2, 4.

Jupiter Communications 1999, 'Travel Suppliers Missing Online Market Potential', http://www.jup.com/company/pressrelease.jsp?doc=pr990512a (visited 25 October 2000)

Kang, Jerry 1998, 'Information Privacy in Cyberspace Transactions', *Stanford Law Review* 50: 1193–294.

Kaplan, B 1967, *An Unhurried View of Copyright*, Columbia University Press, New York.

Kaplan, C. S. 1999, 'On the Web, It's Buyer Beware. But Where?', *New York Times* 26 March. http://www.nytimes.com/library/tech/99/03/cyber/cyberlaw/26law.html (visited 25 October 1999).

Kaplan, David E., and Alec Dubro 1986, *Yakuza: The Explosive Account of Japan's Criminal Underworld*. Collier Books, New York.

Katz, Leo, and James Lindgren 1993, 'Symposium: Blackmail', *University of Pennsylvania Law Review* 141(5): 1565–989.

Kennedy, D. 1996, 'Russian Pleads Guilty to Stealing from Citibank Accounts', http://catless.ncl.ac.uk/Risks/17.61.html#subj4.1 (visited 25 October 2000).

Kitch, E. W. 1977, 'The Nature and Function of the Patent System', *Journal of Law and Economics* 20: 265.

KPMG 1999, *1999 Fraud Survey*, KPMG, Sydney.

Labaton, S. 1999, 'Net Sites Co-opted by Pornographers, *The New York Times On the Web*, 23 September. http://www.grex.org/~tony1/mousetrap.html (visited 25 October 2000)

Lahore, J. 1981, 'The Legal Rationale of the Patent System'. *In Economic Implications of Patents in Australia*, Australian Patent Office, Canberra.

Lamberton, D. M. 1994, 'Innovation and Intellectual Property'. In M. Dodgson and R. Rothwell (eds), *The Handbook of Industrial Innovation*, Edward Elgard, Aldershot, pp. 301–9.

Lanham, D. 1997, *Cross Border Criminal Law*, John Libbey, Sydney.

LaPorta, R., F. Lopez de Silanes, A. Shleifer and R. Vishny 1997, 'Legal Determinants of External Finance', *Journal of Finance* 52(3): 1131–50.

Lash, Alex 1996, 'AOL user guilty of blackmail attempt', CNET News.com, 22 May 1996. http://www.news.cnet.com/news/0-1005-200-311365.html?tag= (visited 15 October 2000).

Lessig, L. 1995, 'The Path Of Cyberlaw', *Yale Law Journal* 104(7): 1743–55.

Lessig, L. 1999, *Code and Other Laws of Cyberspace*, Basic Books, New York.

Levi, M., and J. Handley 1998, *The Prevention of Plastic and Cheque Fraud Revisited*, Home Office Research Study No. 182, Home Office, London.

Levy, S. 1994, 'E-Money (That's What I Want)', *Wired* December: 174–9, 213.

Lindgren, James 1983, 'Blackmail and Extortion'. In Sanford Kadish (ed.), *Encyclopedia of Crime and Justice*, vol. 1, Free Press, New York, pp. 115–19.

Lindsay, N. 2000, 'Jail for internet share vamper', *Australian Financial Review* 31 October.

Lock, Stock and Barrel 2000n *Government Home and Garden Competition.* http://www.lockstockandbarrel.org/Documents/HOUSE-GARDEN.htm (visited 25 October 2000).

Loggin, R. 1720, *The Present Management of the Customs*, J. Roberts, J. Brotherton, E. Curll & J. Fox Printers, London.

Louis Harris & Associates Inc. 1999, *Consumers and the 21st Century: A Survey Conducted for the National Consumers League*, Louis Harris & Associates Inc., New York.

Macdonald, S., and J. Nightingale (eds) 1999, *Information and Organization: A Tribute to the Work of Don Lamberton*, North-Holland, Amsterdam.

MacHarg, M. L., and B. B. Clark (eds) 1999, *International Survey of Investment Adviser Regulation*, 2nd edn, Kluwer Law International, The Hague.

Mackenzie, R. 1998, 'Virtual Money, Vanishing Law: Dematerialisation in electronic funds transfer, financial wrongs and doctrinal makeshifts in English legal structures', *Journal of Money Laundering Control* 2(1): 22–32.

Mackrell, N. 1996, 'Economic Consequences of Money Laundering'. In A. Graycar and P. Grabosky (eds), *Money Laundering in the 21st Century: Risks and Countermeasures*, Australian Institute of Criminology, Canberra, pp. 29–35.

Macy's 2000, 'Macys Home Page'. http://www.macys.com (visited 25 October 2000).

Man Eats Dog 2000, *Switching and Routing in the UK Network.* http://www.members.tripod.com/~iang/med-telefon2.html (visited 24 October 2000).

Markoff, J. 2000, 'An Online Extortion Plot Results in Release of Credit Card Data', *The New York Times on the Web*, 10 January. http://www.nytimes.com/library/tech/00/01/biztech/articles/10hack.html (visited 25 October 2000).

Martinson, J. 1997, 'NASDAQ Move to Curb False Internet Reports', *Financial Times* (London), 5 September, p. 17.

Marx, Gary 1987, 'The Interweaving of Private and Public Police in Undercover Work'. In Clifford Shearing and Philip Stenning (eds), *Private Policing*, Sage Publications, Newbury Park, Ca., pp. 172–93.

Mayer-Schonberger, Viktor 1997, 'Generational Development of Data Protection in Europe'. In Philip Agre and Marc Rotenberg (eds), *Technology and Privacy: the New Landscape*, MIT Press, Cambridge, Mass., pp. 219–41.

McCullagh, A. 1998, 'Legal Aspects of Electronic Contracts and Digital Signatures'. In A. Fitzgerald, B. Fitzgerald, P. Cook and C. Cifuentes (eds), *Going Digital: Legal Issues for Electronic Commerce, Multimedia and the Internet*, Prospect Media, Sydney, pp. 114–26.

Meagher, R. P., W. M. C. Gummow and J. F. R. Lehane 1992, *Equity: Doctrines and Remedies*, 3rd edn, Butterworths, Sydney.

Meijboom, A. P. 1988, 'Problems Related to the Use of EFT and Teleshopping Systems by the Consumer'. In Y. Poullet and G. P. V. Vandenberghe, *Telebanking, Teleshopping and the Law*, Kluwer Law and Taxation Publishers, Deventer, pp. 23–32.

Merry, Sally 1988, 'Legal Pluralism', *Law and Society Review* 22(5): 869–96.

Miller, Greg, and Davan Maharaj 1999, 'N. Hollywood man charged in 1st cyber-stalking case', *Los Angeles Times* 22 January. http://www.cs.csubak.edu/~donna/news/crime.html#stalking (visited 25 October 2000).

Mills, J. 1999, 'Ethics in Governance: Developing Moral Public Service', *Journal of Financial Crime* 7(1): 52–62.

Minidisc 2000 *Music on the Internet.* http://www.hip.atr.co.jp/~eaw/minidisc/music_internet.html (visited 25 October 2000).

Ministerial Council on Consumer Affairs 1999, *The Little Black Book of Scams*, Department of Treasury, Canberra.

Montano, E. 1999, 'Cooperation with the Private Sector in Crime Control'. Paper presented at the Australian Institute of Criminology's Third National Outlook Symposium on Crime in Australia, 'Mapping the Boundaries of Australia's Criminal Justice System', 23 March, Canberra.

Motorola 1999, http://commerce.motorola.com/css/static/mo2000.html (visited 11 May 1999).

Motorola 2000a *Consumer Catalog: Legal terms and Conditions.* http://commerce.motorola.com/consumer/QWhtml/legal_terms.html#disclaimer (visited 25 October 2000).

Motorola 2000b, 'Website'. http://commerce.motorola.com/consumer/QWhtml/privacy.html#child (visited 25 October 2000).

Muldoon, Jennifer, and Melissa Jones 1998, 'Extortion attempt against Qantas Airways'. Paper presented to the Australian Institute of Criminology Conference 'Crime Against Business', 18–19 June, Melbourne. http://www.aic.gov.au/conferences/cab/index.html#18th (visited 25 October 2000).

National Fraud Information Center 2000, *Welcome to the National Fraud Information Center.* http://www.fraud.org (visited 24 July 2000).

Negroponte N. 1995, *Being Digital*, Hodder & Stoughton, London.

Newton, J. 1995, *Organised Plastic Counterfeiting*, HMSO, London.

Newton, T. 1998, 'An Interview with Trevor Newton', *Investing Canada*, 2 February. http://investingcanada.about.com/aboutcanada/investingcanada/library/weekly/1998/aa020298.htm?terms=an+interview+with+trevor+newton (visited 25 October 2000)

New York Stock Exchange 2000, *Regulation.* http://www.nyse.com/regulation/regulation.html (visited 25 October 2000).

Nicholson, E. 1989, 'Hacking away at liberty', *The Times*, 18 April.

NSW Attorney-General's Department 2000, Privacy NSW. http://www.lawlink.nsw.gov.au/pc.nsf/pages/index. (visited 25 October 2000).

NSW Government 2000, http://www.act.gov.au/austouch/austouch.html (visited 19 June 2000).

NSW Parliament 1973, Report on the Law of Privacy (Morrison Report). Government Printer, Sydney.

Oakes, C 1999, 'Password Thief Ransacks AOL', Wired News, 12 October. http://www.wired.com/news/technology/0,1282,31833,00.html (visited 25 October 2000).

O'Brien, Chris 2000, 'The Next Revolution?', *The Age* (Melbourne) IT (2), 27 June, p. 1.

O'Brien, F. 1991, 'Sentencing and Social Security Fraud', *Law Institute Journal*, June, pp. 519–21.

Office of the Commonwealth Privacy Commissioner 2000, *Guidelines on Workplace E-mail, Web Browsing and Privacy*, 30 March. http://www.privacy.gov.au/issues/p7_4.html (visited 25 October 2000).

Office of Fair Trading 1998, 'Internet Scams Deleted, Sweep Identifies "Get Rich Quick" Schemes', *Fair Trading Magazine*, Spring, Office of Fair Trading, London.

Office of Government Information Technology 1998, *Gatekeeper: A Strategy for Public Key Technology Use in the Government*, AGPS, Canberra.

Office of Strategic Crime Assessments and Victoria Police 1997, *Computer Crime and Security Survey*, Attorney-General's Department, Canberra.

OECD (Organization for Economic Co-operation and Development) 1986, *Computer-Related Crime: Analysis of Legal Policy*, OECD, Paris.

Osborne, D., and T. Gaebler 1992, *Reinventing Government*, Addison Wesley, Boston.

O'Sullivan, J., and T. Damian 1999, 'International Survey of Investment Adviser Regulation: Australia'. In M. L. MacHarg and B. B. Clark (eds) 1999, *International Survey of Investment Adviser Regulation*, 2nd edn, Kluwer Law International, The Hague, pp. 33–60.

O'Toole, G. J. A. 1978, *The Private Sector: Private Spies, Rent-A-Cops, and the Police-Industrial Complex*, W. W. Norton & Co. Inc., New York.

Parker, Donn B. 1976, *Crime by Computer*, Charles Scribner's Sons, New York.

Pauley, G. 1999, *Gerry Pauley's Australian Stocks and Shares* http://users.wantree.com.au/~tpauleyg/shares.html (visited 25 October 2000).

Peachey, D., and J. Blau 1995, 'Fraud Spreading in Corporate Nets', *Communications Week International* 137: 1.

Pearson, G. 1996, 'Naked Men, Food and Water: Marketing Law and Codes of Practice', *Current Commercial Law* 4(1): 21–32.

Pelline, J. 1997, 'Hacker Admits to AOL Piracy', *CNET News.* http://news.cnet.com/news/0-1005-200-315653.html (visited 25 October 2000).

Perry, S. 1995, *Economic Espionage and Corporate Responsibility* http://www.the-south.com/USInternational/econ.html (visited 25 October 2000).

Petersen, H., and G. Poupard 1997, 'Efficient scalable fair cash with off-line extortion prevention', *Technical Report LIENS-97-7*, Ecole Normale Superieure, Paris.

Phillipps, T. 1999, 'Internet Fraud and the Financial Services Sector'. Paper presented to the 1999 Internet Fraud Summit, 19 April, Sydney.

Phone Losers of America 2000, http://www.phonelosers.org/ (visited 25 October 2000).

Picciotto, S. 1997, 'The Regulatory Criss-Cross: Interaction between Jurisdictions and The Construction of Global Regulatory Networks'. In W. Bratton, J. McCathery and S. Picciotto (eds), *International Regulatory Competition and Coordination: Perspectives in Economic Regulation in Europe and the United States*, Oxford University Press, New York.

Platform for Privacy Preferences 1999, http://www.w3.org/P3P (visited 25 October 2000).

Preiss, R. T. 1998, 'The Consequences of Anonymous Access to the Financial Payments System', *Journal of Money Laundering Control* 2(1): 7–13.

Prentice, R. A. 1998, 'The Future of Corporate Disclosure: The Internet, securities fraud, and rule 10b-5', *Emory Law Journal* 47(1): 1-88. http://www.law.emory.edu/ELJ/volumes/win98/prentice.html (visited: 25 October 2000).

Price, G., and C. Fox 1997, 'The Massachusetts Bank Reporting Project: An Edge Against Elder Financial Exploitation', *Journal of Elder Abuse and Neglect* 8(4): 59–71.

Raab, Charles D. 1997, 'Co-producing Data Protection', *International Review of Law Computers and Technology* 11(1): 11–24.

Rand Corporation 1999, *Economic Costs and Implications of High-Technology Hardware Theft*, Rand Corporation, Santa Monica, California.

Rastan, C. 1996, 'Not So Funny Money: Curbing the Counterfeiters', *Crime Prevention News* March, pp. 17–19.

Rawitch, R. 1979, 'Expected Bank Plot to Fail', *Los Angeles Times*, 23 February, pp. 1, 27.

Research Group into the Law Enforcement Implications of Electronic Commerce 1999, *Issues Paper: Series No 1*, Attorney-General's Department, Canberra.

Reuter, Peter 1983, *Disorganized Crime: The Economics of the Visible Hand*, MIT Press, Cambridge, Mass.

Reuters 1998, Japan reports cyber bank heist. 5 January 1998. http://www3.zdnet.com/zdnn/content/reut/0105/268074.html (visited 25 October 2000).

Reuters 1999, 'Brit cops post informants list' Reuters Wired News, 26 February 1999. http://www.wired.com/news/news/culture/story/18147.html (visited 25 October 2000).

Richtel, M. 1999a, 'Super-Fast Computer Virus Heads Into the Workweek', *New York Times*, 29 March. http://nytimes.com/library/tech/99/03/biztech/articles/29virus.html (visited 29 March 1999).

Richtel, M. 1999b, 'Investigators Face a Glut of Confiscated Computers', *New York Times*, 27 August 1999. http://www.nytimes.com/library/tech/99/08/cyber/cyberlaw/27law.html (visited 25 October 2000).

Ricketson, M., and S. Richardson 1998, *Intellectual Property: Cases, commentary and materials*, 2nd edn, Butterworths, Sydney.

Robb, G. 1992, *White-collar Crime in Modern England: Financial Fraud and Business Morality, 1845–1929*, Cambridge University Press, Cambridge.

Robertson, R. A. 1996, 'Personal Investing in Cyberspace and the Federal Securities Laws', *Securities Regulation Law Journal* 23: 347–422.

Robinson, J. 1999, 'Remarks Prepared for Delivery to the American Bar Association', National Institute on White-Collar Crime, San Francisco.

Rochlin, Gene I. 1997, *Trapped in the Net: The Unanticipated Consequences of Computerization*, Princeton University Press, Princeton.

Rohm, Wendy Goldman 1998, 'Microsoft Emails obtained by Justice suggest a pattern of misbehavior', *The Red Herring Magazine*, November. http://www.redherring.com/mag/issue60/microsoft.html (visited 25 October 2000).

Romei, S. 1999, 'In Cyberspace Scams, You Can't Hear Suckers Scream', *Australian*, 25 August, p. 26.

Rose, N., and P. Miller 1992, 'Political Power Beyond the State: Problematics of government', *British Journal of Sociology* 43: 173–205.

Rosen, Jeffrey 2000a, 'The Eroded Self', *New York Times Magazine* 30 April.

Rosen, Jeffrey 2000b, *The Unwanted Gaze: The Destruction of Privacy in America*, Random House, New York.

Rosoff, S. M., H. N. Pontell and R. Tillman 1998, *Profit Without Honor: White Collar Crime and the Looting of America*, Prentice Hall Inc., Upper Saddle River, New Jersey.

Rotenberg, M. 1998, Statement before the Subcommittee on Courts and Intellectual Property, Committee on the Judiciary, US House of Representatives, March 26, 1998. http://www.house.gov/judiciary/41180.htm (visited 25 October 2000).

Rothchild, J. 1999, 'Protecting the Digital Consumer: The Limits of Cyberspace Utopianism', *Indiana Law Journal* 74: 893–989.

Ryan, Nicholas 1997, 'Happy Hardcore'. http://www.yaleherald.com/archive/xxiv/10.3.97/exclusive/letter.html (visited 25 October 2000).

Sacramento Bee 2000, Letter Sent to the *New York Times*. http://www.unabombertrial.com/manifesto/nytletter.html (visited 25 October 2000).

Sakurada, Keiji 1994, 'Yakuza Storming the Corporate Ship', *Tokyo Business* April: 46–7.

Salmons, R. 2000, 'Cut-Price War Hits E*Trade', *The Age* (Melbourne) 4 April, p. 1.

Samuelson P. 1996, 'Authors' Rights in Cyberspace: Are New International Rules Needed?', *First Monday* 1 (October), at http://www.firstmonday.dk/issues/issue4/samuelson/ (visited 24 October 2000).

San Jose Mercury News 1999, 'Intel to change Pentium III due to privacy concerns', *San Jose Mercury News* 25 January. http://spyglass1.sjmercury.com/breaking/docs/062233.htm (visited 24 June 1999).

Scannell, Ed 1999, 'IT gets tools to thwart Webjackers'. http://www.infoworld.com/cgi-bin/displayStory.pl?991029.hnwebjack.htm (visited 25 October 2000).

Schwartau, W. 1999, 'Cyber-vigilantes hunt down Hackers', CNN.com, 12 January. http://cnn.com/TECH/computing/9901/12/cybervigilantes.idg/ (visited 25 October 2000).

Schwartz, Paul 1997, 'Privacy and the Economics of Personal Health Care Information', *Texas Law Review* 76(1): 1–75.

Sealy, L. S. 1962, 'Fiduciary Relationships', *Cambridge Law Journal* April: 69–81.

SEC (US Securities and Exchange Commission) 1998a, *Statement of the Commission Regarding Use of Internet Web Sites to Offer Securities, Solicit Securities Transactions or Advertise Investment Services Offshore*, Release Nos. 33-7516, 34-39779, IA-1710, IC-23071, International Series Release No. 1125, 23 March.

SEC (US Securities and Exchange Commission) 1998b, *Internet Fraud: How to Avoid Internet Investment Scams*. http://www.sec.gov/consumer/cyberfr.htm (visited 23 July 2000).

SEC (US Securities and Exchange Commission) 2000 *SEC Division of Enforcement Complaint Center*, http://www.sec.gov/enforce/comctr.htm (visited 25 October 2000).

Serious Fraud Office 1997, *Annual Report 1996/1997*, HMSO, London. http://www.sfo.gov.uk/anrep97.htm (visited 25 October 2000).

Serious Fraud Office 1998, *Annual Report 1997/1998*, HMSO, London. http://www.sfo.gov.uk/anrep98.htm (visited 25 October 2000).

Shapiro, Susan 1987, 'Policing Trust'. In Shearing and Stenning (eds), *Private Policing*, Sage Publications, Newbury Park, pp. 194–220.

Shearing, C. 1993, 'A Constitutive Conception of Regulation'. In P. N. Grabosky and J. Braithwaite (eds), *Business Regulation and Australia's Future*, Australian Institute of Criminology, Canberra, pp. 67–80.

Sheehan, B. 1999, *billsheehan.com*. http://billsheehan.com (visited 14 June 1999).

Shimomura, T. (with John Markoff) 1996, *Takedown*, Hyperion, New York.

Sibert, O., D. Bernstein and D. Van Wie 1996, *Securing the Content, Not the Wire, for Information Commerce*. http://www.star-lab.com/secure-the-content.html (visited 25 October 2000).

Sieber, Ulrich 1988, *Computer Crime and Criminal Information Law – New Trends in the International Risk and Information Society*. http://www.jura.uni-wuerzburg.de/lst/sieber/mitis/ComCriCriInf.htm#fn0 (visited 25 October 2000).

Silverman, J. 1997, 'Liability in Cyberspace II: Disclosure and Securities Fraud on the Internet', *Computer Law and Security Report* 13(6): 447–50.

Simpson, A. F. 1997, '"Bits" of Disclosure: Communications Technology and Securities Regulation in Australia', *Journal of International Banking Law* 12: 371–82.

Simpson, A. F. 1998, 'Securities Regulation for the Information Age'. In G. Walker, B. Fisse and I. Ramsay (eds), *Securities Regulation in Australia and New Zealand*, 2nd edn, LBC Information Services, Sydney, pp. 33–54.

Sinclair, J. 1999a, 'Sex Sites and the gov.au Connection', *The Age* (Melbourne) 8 June, p. IT-3.

Sinclair, J. 1999b, 'Tax Office Cracks Down on Illegal Operators', *The Age* (Melbourne) IT, 23 February, pp. IT-1–2.

Smith, R. G. 1997, 'Plastic Card Fraud', in *Trends and Issues in Crime and Criminal Justice*, No. 71, Australian Institute of Criminology, Canberra.

Smith, R. G. 1998a, 'Best Practice in Fraud Prevention'. In *Trends and Issues in Crime and Criminal Justice*, No. 100, Australian Institute of Criminology, Canberra.

Smith, R. G. 1998b, 'Plastic Card Fraud', *Australian Banker* 112(3): 92–9.

Smith, R. G. 1998c, 'The Regulation of Telemedicine'. In R. G. Smith (ed.), *Health Care, Crime and Regulatory Control*, Hawkins Press, Sydney, pp. 190–203.

Smith, R. G. 1999a, 'Electronic Medicare Fraud: Current and Future Risks'. In *Trends and Issues in Crime and Criminal Justice*, No. 114, Australian Institute of Criminology, Canberra.

Smith, R. G. 1999b, 'Identity-Related Economic Crime: Risks and Countermeasures'. In *Trends and Issues in Crime and Criminal Justice*, No. 129, Australian Institute of Criminology, Canberra.

Smith, R. G., M. N. Holmes and P. Kaufmann 1999, 'Nigerian Advance Fee Fraud'. In *Trends and Issues in Crime and Criminal Justice*, Australian Institute of Criminology, Canberra.

Sneddon, M. 1998, 'Legislating to Facilitate Electronic Signatures and Records: Exceptions, Standards and the Impact on the Statute Book', *University of New South Wales Law Journal* 21 (2): 334–403.

Sommer, Peter 1999 *Cyber Extortion.* http://www.westcoast.com/securecomputing/april/extortion/ (visited 27 November 2000).

Sparrow, M. K. 1996, *License to Steal: Why Fraud Plagues America's Health Care System*, Westview Press, Colorado.

Spice, Linda, and Lisa Sink 1999, 'Criminal charges sought over posting of nude photos on Web', *Milwaukee Journal Sentinel*, 20 May. http://www.jsonline.com/news/metro/apr99/990520criminalchargessought.asp (visited 25 October 2000).

Spinks, Peter 1996, 'Tests Show Up Smart Card Flaws', *The Age* (Melbourne) 6 December.

Spink, Paul 1997, 'Misuse of police computers', *Juridical Review* 1997: 219–29.

Starek, R. B., and L. M. Rozell 1997, 'The Federal Trade Commission's Commitment to On-Line Consumer Protection', *Journal of Computer and Information Law* 15: 679–702.

Steele, R. D. 1993, *Theory and Practice of Intelligence in the Age of Information.* http://www.oss.net/Papers/training/Lesson002Handout.html (visited 4 November 2000).

Stefik M. 1997, 'Shifting The Possible: How Trusted Systems And Digital Property Rights Challenge Us To Rethink Digital Publishing', *Berkeley Technology Law Journal* 12(1), at http://www.law.berkeley.edu/journals/btlj/articles/12-1/stefik.html (visited 25 October 2000).

Stirland, S. 1996, 'News and Trends: Securities Regulators Prowl the Net, Looking for Lawbreakers', *Bond Buyer* 13 November, p. 34.

Stoll, C. 1991, *The Cuckoo's Egg*, Pan Books, London.

Streeck, Wolfgang, and Philippe Schmitter 1985, 'Community, Market, State and Association? The prospective contribution of Interest Governance to Social Order'. In Streeck and Schmitter (eds), *Private Interest Government: Beyond Market and State*, Sage Publications, Beverly Hills, pp. 1–29.

Sullivan, C. 1987, 'Unauthorised Automatic Teller Machine Transactions: Consequences for Customers of Financial Institution', *Australian Business Law Review* 15(3): 187–214.

Svensson, N. 1997, 'Swedish Secret Police Investigates Espionage Against Ericsson Telecom', *Expressen*, Sweden, March.

Sydney Morning Herald 2000, 'Internet ban to protect Jews: Orthodox rabbis believe the Internet is a Trojan Horse for secular filth'. http://www.smh.com.au/news/0001/10/world/world1.html (visited 25 October 2000).

Sykes, T. 1994, *The Bold Riders: Behind Australia's Corporate Collapses*, Allen & Unwin, Sydney.

Taylor, B. 2000, 'Prairie Dog: The Cutest Internet Watchdog'. http://www.prairie-dog.net/ (visited 25 October 2000).

Tedeschi, Bob 1999, 'Net Companies Look Offline for Consumer Data', New York Times on the Web, 21 June. http://www.nytimes.com/library/tech/99/06/cyber/commerce/21commerce.html (visited 25 October 2000).

Tendler, S., and N. Nuttall 1996, 'Hackers Leave Red-Faced Yard with $1.29m Bill', *The Australian*, 6 August, p. 37a.

Teubner, G. 1983, 'Substantive and Reflexive Elements in Modern Law', *Law and Society Review* 17: 239–86.

The Australian 2000 'Beware, now Big Brother is online', *The Australian* 26 August 2000, p. 48.

The Business UK 1999, 'Cyber Scams' http://www.thebusinessuk.com/fraud/cyber.htm (visited 18 December 1999).

Thomas-Lester, Avis, and Toni Locy 1997, 'Chief's Friend Accused of Extortion' *Washington Post*, 26 November, Page A01.

Tomasic, R. 1991, *Casino Capitalism: Insider Trading in Australia*, Australian Institute of Criminology, Canberra.

TRUSTe 2000, *Building a Web you can believe in* http://www.truste.org (visited 25 October 2000).

Turner, A. E. 1962, *The Law of Trade Secrets*, Law Book Company, Sydney.

Tweney, D. 1998, 'Sex Scam Points Out Lack of Safeguards in Online Business', 3 August 1998. http://tweney.com/prophet/980803prophet.htm (visited 25 October 2000).

Tynan, D. 1999, 'Web Scam: How One ISP Took the Money and Ran'. http://cnn.com/TECH/computing/9904/28/scam.idg/ (visited 25 October 2000).

UK National Audit Office 1994, *HM Customs and Excise: Prevention and Detection of Internal Fraud*, HC 244, HMSO, London.

UK Office of Fair Trading 2000, *Protecting Consumers*. http://www.oft.gov.uk (visited 25 October 2000).

UK Prime Minister 1999, http://www.number-10.gov.uk/public/info/releases/publications/infoagefeat.htm (visited 19 June 1999).

UNCITRAL (UN Commission on International Trade Law) 1996, *Model Law on Electronic Commerce*. http://www.un.or.at/uncitral/texts/electcom/ml-ec.htm (visited 25 October 2000).

United Kingdom 2000, *Regulation of Investigatory Powers Bill*. http://www.parliament.the-stationery-office.co.uk/pa/ld199900/ldbills/061/en/00061x–.htm (visited 25 October 2000).

US Department of Justice 2000, *Internet Fraud: Appendix B*, Report of the Criminal Division's Computer Crime and Intellectual Property Section. http://www.cybercrime.gov/append.htm (visited 25 October 2000).

Unlistedphonenumbers.com 2000 http://www.unlistedphonenumbers.com/index.htm (visited 25 October 2000).

Upton, Jody 1999, 'U-M medical records end up on Web', *Detroit News* 12 February. http://detnews.com/1999/metro/9902/12/02120114.htm (visited 25 October 2000).

Varney, C. 1996, 'Regulating Cyberspace: An Off the Record Interview with Federal Trade Commissioner Christine Varney', *Computer Underground Digest* 8(24).

Vaver, P. F. 1979, 'Trade Secrets: A Commonwealth Perspective', *European Intellectual Property Review* 1: 301.

Victoria Police and Deloitte Touche Tohmatsu 1999, *Computer Crime and Security Survey*, Victoria Police Computer Crime Squad and Deloitte Touche Tohmatsu, Melbourne.

Victorian Government 2000, http://maxi.com.au (visited 25 October 2000).

Visa International 2000, 'The World of Visa Secure Electronic Commerce'. http://www-s2.visa.com/nt/sec/no_shock/entry_L.html

Wahlert, G. 1998, 'Crime in Cyberspace: Trends in Computer Crime in Australia', *Platypus Magazine: Journal of the Australian Federal Police* 59: 3–9.

Wallace, R. 1999, 'Conducting a Fraud Investigation Within a Telecommunications Company', Vodaphone Targeting Fraud ICM Conference, Melbourne, 15–16 September.

Waller, D. 1992a, 'Truth in Advertising', *City Ethics* 6(Summer): 4–5.

Waller, D. 1992b, 'Truth in Advertising II', *City Ethics* 7(Autumn): 4.

Walsh, L. 1993, *Final Report of the Independent Counsel for Iran/Contra Matters, vol. 1: Investigations and Prosecutions*. Lawrence E. Walsh, Independent Counsel. August 4, 1993. Washington, DC United States Court of Appeals for the District of Columbia Circuit, Division for the Purpose of Appointing Independent Counsel Division No. 86-6. http://www.fas.org/irp/offdocs/walsh/part_iii.htm (visited 25 October 2000).

Warren, Samuel, and Louis Brandeis 1890, 'The Right to Privacy', *Harvard Law Review* 4(5): 193–220.

Warton, A. 1999, 'Electronic Benefit Transfer Fraud: The Challenge for Federal Law Enforcement', *Platypus Magazine: The Journal of the Australian Federal Police* 65 (December): 38–44.

Webb, B. 1996, 'Preventing Plastic Card Fraud in the UK', *Security Journal* 7: 23–5.

Wertheimer, Alan 1987, *Coercion*, Princeton University Press, Princeton.

Westin, Alan 1967, *Privacy and Freedom*, Athenaeum, New York.

Whitaker, Reg 1999, *The End of Privacy: How Total Surveillance is Becoming a Reality*, New Press, New York.

Whitford, D., and R. Rao 1998, 'Trade Fast, Trade Cheap'. http://library.northernlight.com/SG19990714120000591.html?cb=13&sc=0#doc (visited 25 October 2000).

Wilke, M. 1993, *Women Social Security Offenders: Experiences of the Criminal Justice System in Western Australia*, University of Western Australia, Crime Research Centre, Perth.

Williams, Glanville 1954, 'Blackmail', *Criminal Law Review* 79–92, 162–72, 240–6.

WIPO (World Intellectual Property Organization) 1996, 'WIPO Copyright Treaty', WIPO, Geneva, 20 December 1996. http://www.wipo.org/eng/diplconf/distrib/94dc.htm (visited 25 October 2000).

Wired News 1998a, Internet Hacking for Dummies. http://www.wired.com/news/technology/0,1282,10459,00.html (visited 25 October 2000).

Wired News 1998b, 'Ex-Kodak Worker Charged in Theft' 9 July. http://www.wired.com/news/politics/0,1283,13581,00.html (visited 25 October 2000).

Wolverton, Troy 1999a, 'Yahoo plugs security breach' CNET News.com, 8/4/99. http://www.news.com/News/Item/0,4,34841,00.html (visited 25 October 2000).

Wolverton, Troy 1999b, Reference: 'AT&T reveals 1,800 customer email addresses', CNET News.com, 15/4/99. http://www.news.com/News/Item/0,4,35225,00.html (visited 25 October 2000).

Wolverton, Troy 1999c, 'Nissan privacy goof exposes e-mail addresses', CNET News.com, 15 April. http://www.news.com/News/Item/0,4,35168,00.html (visited 25 October 2000).

Wraith, J. 1999, 'Telecommunications Fraud in the Late 90s', *INTERSEC* 9(10): 307–10.

Yamey, B.S. 1970, 'Monopoly, Competition and the Incentive to Invent: a Comment', *Journal of Law and Economics* 8: 253.

Young, Adam, and Moti Yung 1996, 'Cryptovirology: Extortion-Based Security Threats and Countermeasures', *Proceedings of the 1996 IEEE Symposium on Research in Security and Privacy* (SECURE '96).

Young, S. 1999, 'Thumbs Up for Fingerprint-Based Ids', *The Age* (Melbourne) IT: p. 4.

Zanin, B. 1998, 'Helping Seniors Help Themselves', *RCMP Gazette* 60(11): 26.

Ziegelaar, M. 1998, 'Insider Trading Law in Australia'. In G. Walker, B. Fisse and I. Ramsay (eds), *Securities Regulation in Australia and New Zealand*, 2nd edn, LBC Information Services, Sydney, pp. 556–94.

Index

www.ingramcontent.com/pod-product-compliance
Ingram Content Group UK Ltd.
Pitfield, Milton Keynes, MK11 3LW, UK
UKHW050112180125
453697UK00008B/148